Finite Markov Chains and Algorithmic Applications

Finite Markov Chains
and
Algorithmic Applications

Olle Häggström

Matematisk statistik, Chalmers tekniska högskola och Göteborgs universitet

CAMBRIDGE
UNIVERSITY PRESS

CAMBRIDGE UNIVERSITY PRESS
Cambridge, New York, Melbourne, Madrid, Cape Town,
Singapore, São Paulo, Delhi, Tokyo, Mexico City

Cambridge University Press
The Edinburgh Building, Cambridge CB2 8RU, UK

Published in the United States of America by
Cambridge University Press, New York

www.cambridge.org
Information on this title: www.cambridge.org/9780521890014

First published 2002
Fifth printing 2008

A catalogue record for this publication is available from the British Library

Library of Congress cataloguing in publication data

ISBN 978-0-521-81357-0 Hardback
ISBN 978-0-521-89001-4 Paperback

Contents

Preface

The first version of these lecture notes was composed for a last-year under-graduate course at Chalmers University of Technology, in the spring semester 2000. I wrote a revised and expanded version for the same course one year later. This is the third and final (?) version.

The notes are intended to be sufficiently self-contained that they can be read without any supplementary material, by anyone who has previously taken (and passed) some basic course in probability or mathematical statistics, plus some introductory course in computer programming.

The core material falls naturally into two parts: Chapters 2–6 on the basic theory of Markov chains, and Chapters 7–13 on applications to a number of randomized algorithms.

Markov chains are a class of random processes exhibiting a certain "mem-oryless property", and the study of these – sometimes referred to as Markov theory – is one of the main areas in modern probability theory. This area cannot be avoided by a student aiming at learning how to design and implement randomized algorithms, because Markov chains are a fundamental ingredient in the study of such algorithms. In fact, any randomized algorithm can (often fruitfully) be viewed as a Markov chain.

I have chosen to restrict the discussion to discrete time Markov chains with finite state space. One reason for doing so is that several of the most important ideas and concepts in Markov theory arise already in this setting; these ideas are more digestible when they are not obscured by the additional technicalities arising from continuous time and more general state spaces. It can also be argued that the setting with discrete time and finite state space is the most natural when the ultimate goal is to construct algorithms: Discrete time is natural because computer programs operate in discrete steps. Finite state space is natural because of the mere fact that a computer has a finite amount of memory, and therefore can only be in a finite number of distinct "states". Hence, the Markov chain corresponding to a randomized algorithm implemented on a real computer has finite state space.

However, I do not claim that more general Markov chains are irrelevant to the study of randomized algorithms. For instance, an infinite state space is sometimes useful as an approximation to (and easier to analyze than) a finite but very large state space. For students wishing to dig into the more general Markov theory, the final chapter provides several suggestions for further reading.

Randomized algorithms are simply algorithms that make use of random number generators. In Chapters 7–13, the Markov theory developed in previous chapters is applied to some specific randomized algorithms. The Markov chain Monte Carlo (MCMC) method, studied in Chapters 7 and 8, is a class of algorithms which provides one of the currently most popular methods for simulating complicated stochastic systems. In Chapter 9, MCMC is applied to the problem of counting the number of objects in a complicated combinatorial set. Then, in Chapters 10–12, we study a recent improvement of standard MCMC, known as the Propp–Wilson algorithm. Finally, Chapter 13 deals with simulated annealing, which is a widely used randomized algorithm for various optimization problems.

It should be noted that the set of algorithms studied in Chapters 7–13 constitutes only a small (and not particularly representative) fraction of all randomized algorithms. For a broader view of the wide variety of applications of randomization in algorithms, consult some of the suggestions for further reading in Chapter 14.

The following diagram shows the structure of (essential) interdependence between Chapters 2–13.

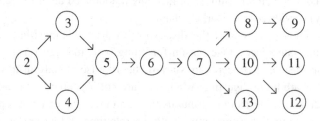

How the chapters depend on each other.

Regarding exercises: Most chapters end with a number of problems. These are of greatly varying difficulty. To guide the student in the choice of problems to work on, and the amount of time to invest into solving the problems, each problem has been equipped with a parenthesized number between (1) and

(10) to rank the approximate size and difficulty of the problem. (1) means
that the problem amounts simply to checking some definition in the chapter (or
something similar), and should be doable in a couple of minutes. At the other
end of the scale, (10) means that the problem requires a deep understanding
of the material presented in the chapter, and at least several hours of work.
Some of the problems require a bit of programming; this is indicated by an
asterisk, as in (7*).

□ □ □ □

I am grateful to Sven Erick Alm, Nisse Dohrnér, Devdatt Dubhashi, Mihyun
Kang, Dan Mattsson, Jesper Møller and Jeff Steif, who all provided corrections
to and constructive criticism of earlier versions of this manuscript.

1

Basics of probability theory

The majority of readers will probably be best off by taking the following piece of advice:

Skip this chapter!

Those readers who have previously taken a basic course in probability or mathematical statistics will already know everything in this chapter, and should move right on to Chapter 2. On the other hand, those readers who lack such background will have little or no use for the telegraphic exposition given here, and should instead consult some introductory text on probability. Rather than being read, the present chapter is intended to be a collection of (mostly) definitions, that can be consulted if anything that looks unfamiliar happens to appear in the coming chapters.

□ □ □ □

Let Ω be any set, and let Σ be some appropriate class of subsets of Ω, satisfying certain assumptions that we do not go further into (closedness under certain basic set operations). Elements of Σ are called **events**. For $A \subseteq \Omega$, we write A^c for the **complement** of A in Ω, meaning that

$$A^c = \{s \in \Omega : s \notin A\}.$$

A **probability measure** on Ω is a function $\mathbf{P} : \Sigma \to [0, 1]$, satisfying

(i) $\mathbf{P}(\emptyset) = 0$.
(ii) $\mathbf{P}(A^c) = 1 - \mathbf{P}(A)$ for every event A.
(iii) If A and B are disjoint events (meaning that $A \cap B = \emptyset$), then $\mathbf{P}(A \cup B) = \mathbf{P}(A) + \mathbf{P}(B)$. More generally, if $A_1, A_2, \ldots$ is a countable sequence

of disjoint events ($A_i \cap A_j = \emptyset$ for all $i \neq j$), then $\mathbf{P}\left(\bigcup_{i=1}^{\infty} A_i\right) = \sum_{i=1}^{\infty} \mathbf{P}(A_i)$.

Note that (i) and (ii) together imply that $\mathbf{P}(\Omega) = 1$.

If A and B are events, and $\mathbf{P}(B) > 0$, then we define the **conditional probability of A given B**, denoted $\mathbf{P}(A \mid B)$, as

$$\mathbf{P}(A \mid B) = \frac{\mathbf{P}(A \cap B)}{\mathbf{P}(B)}.$$

The intuitive interpretation of $\mathbf{P}(A \mid B)$ is as how likely we consider the event A to be, given that we know that the event B has happened.

Two events A and B are said to be **independent** if $\mathbf{P}(A \cap B) = \mathbf{P}(A)\mathbf{P}(B)$. More generally, the events $A_1, \ldots, A_k$ are said to be independent if for any $l \leq k$ and any $i_1, \ldots, i_l \in \{1, \ldots, k\}$ with $i_1 < i_2 < \cdots < i_l$ we have

$$\mathbf{P}\left(A_{i_1} \cap A_{i_2} \cap \cdots \cap A_{i_l}\right) = \prod_{n=1}^{l} \mathbf{P}(A_{i_n}).$$

For an infinite sequence of events $(A_1, A_2, \ldots)$, we say that $A_1, A_2, \ldots$ are independent if $A_1, \ldots, A_k$ are independent for any k.

Note that if $\mathbf{P}(B) > 0$, then independence between A and B is equivalent to having $\mathbf{P}(A \mid B) = \mathbf{P}(A)$, meaning intuitively that the occurrence of B does not affect the likelihood of A.

A **random variable** should be thought of as some random quantity which depends on chance. Usually a random variable is real-valued, in which case it is a function $X : \Omega \rightarrow \mathbf{R}$. We will, however, also consider random variables in a more general sense, allowing them to be functions $X : \Omega \rightarrow S$, where S can be any set.

An event A is said to be **defined in terms of the random variable X** if we can read off whether or not A has happened from the value of X. Examples of events defined in terms of the random variable X are

$$A = \{X \leq 4.7\} = \{\omega \in \Omega : X(\omega) \leq 4.7\}$$

and

$$B = \{X \text{ is an even integer}\}.$$

Two random variables are said to be independent if it is the case that whenever the event A is defined in terms of X, and the event B is defined in terms of Y, then A and B are independent. If $X_1, \ldots, X_k$ are random variables, then they are said to be independent if $A_1, \ldots, A_k$ are independent whenever each A_i is defined in terms of X_i. The extension to infinite sequences is similar: The random variables $X_1, X_2, \ldots$ are said to be independent if for any sequence

$A_1, A_2, \ldots$ of events such that for each i, A_i is defined in terms of X_i, we have that $A_1, A_2, \ldots$ are independent.

A distribution is the same thing as a probability measure. If X is a real-valued random variable, then the **distribution** μ_X of X is the probability measure on **R** satisfying $\mu_X(A) = \mathbf{P}(X \in A)$ for all (appropriate) $A \subseteq \mathbf{R}$. The distribution of a real-valued random variable is characterized in terms of its **distribution function** $F_X : \mathbf{R} \to [0, 1]$ defined by $F_X(x) = \mathbf{P}(X \le x)$ for all $x \in \mathbf{R}$.

A distribution μ on a finite set $S = \{s_1, \ldots, s_k\}$ is often represented as a vector $(\mu_1, \ldots, \mu_k)$, where $\mu_i = \mu(s_i)$. By the definition of a probability measure, we then have that $\mu_i \in [0, 1]$ for each i, and that $\sum_{i=1}^{k} \mu_i = 1$.

A sequence of random variables $X_1, X_2, \ldots$ is said to be **i.i.d.**, which is short for **independent and identically distributed**, if the random variables

(i) are independent, and

(ii) have the same distribution function, i.e., $\mathbf{P}(X_i \le x) = \mathbf{P}(X_j \le x)$ for all i, j and x.

Very often, a sequence $(X_1, X_2, \ldots)$ is interpreted as the evolution in time of some random quantity: X_n is the quantity at time n. Such a sequence is then called a **random process** (or, sometimes, **stochastic process**). Markov chains, to be introduced in the next chapter, are a special class of random processes.

We shall only be dealing with two kinds of real-valued random variables: **discrete** and **continuous** random variables. The discrete ones take their values in some finite or countable subset of **R**; in all our applications this subset is (or is contained in) $\{0, 1, 2, \ldots\}$, in which case we say that they are **nonnegative integer-valued** discrete random variables.

A **continuous** random variable X is a random variable for which there exists a so-called **density function** $f_X : \mathbf{R} \to [0, \infty)$ such that

$$\int_{-\infty}^{x} f_X(x)dx = F_X(x) = \mathbf{P}(X \le x)$$

for all $x \in \mathbf{R}$. A very well-known example of a continuous random variable X arises by letting X have the Gaussian density function $f_X(x) = \frac{1}{\sqrt{2\pi\sigma^2}}e^{-((x-\mu)^2)/2\sigma^2}$ with parameters μ and $\sigma > 0$. However, the only continuous random variables that will be considered in this text are the **uniform** $[0, 1]$ ones, which have density function

$$f_X(x) = \left\{ \begin{array}{ll} 1 & \text{if } x \in [0, 1] \\ 0 & \text{otherwise} \end{array} \right.$$

and distribution function

$$F_X(x) = \int_{-\infty}^{x} f_X(x)dx = \begin{cases} 0 & \text{if } x \le 0 \\ x & \text{if } x \in [0, 1] \\ 1 & \text{if } x \ge 1. \end{cases}$$

Intuitively, if X is a uniform $[0, 1]$ random variable, then X is equally likely to take its value anywhere in the unit interval $[0, 1]$. More precisely, for every interval I of length a inside $[0, 1]$, we have $\mathbf{P}(X \in I) = a$.

The **expectation** (or **expected value**, or **mean**) $\mathbf{E}[X]$ of a real-valued random variable X is, in some sense, the "average" value we expect from x. If X is a continuous random variable with density function $f_X(x)$, then its expectation is defined as

$$\mathbf{E}[X] = \int_{-\infty}^{\infty} x f_X(x)dx$$

which in the case where X is uniform $[0, 1]$ reduces to

$$\mathbf{E}[X] = \int_{0}^{1} x \, dx = \frac{1}{2}.$$

For the case where X is a nonnegative integer-valued random variable, the expectation is defined as

$$\mathbf{E}[X] = \sum_{k=1}^{\infty} k\mathbf{P}(X = k).$$

This can be shown to be equivalent to the alternative formula

$$\mathbf{E}[X] = \sum_{k=1}^{\infty} \mathbf{P}(X \ge k). \tag{1}$$

It is important to understand that the expectation $\mathbf{E}[X]$ of a random variable can be infinite, even if X itself only takes finite values. A famous example is the following.

> **Example 1.1: The St Petersburg paradox.** Consider the following game. A fair coin is tossed repeatedly until the first time that it comes up tails. Let X be the (random) number of heads that come up before the first occurrence of tails. Suppose that the bank pays 2^X roubles depending on X. How much would you be willing to pay to enter this game?
>
> According to the classical theory of hazard games, you should agree to pay up to $\mathbf{E}[Y]$, where $Y = 2^X$ is the amount that you receive from the bank at the end of the game. So let's calculate $\mathbf{E}[Y]$. We have
>
> $$\mathbf{P}(X = n) = \mathbf{P}(n \text{ heads followed by 1 tail}) = \left(\frac{1}{2}\right)^{n+1}$$

for each n, so that

$$
\begin{aligned}
\mathbf{E}[Y] &= \sum_{k=1}^{\infty} k\mathbf{P}(Y=k) = \sum_{n=0}^{\infty} 2^n \mathbf{P}(Y=2^n) \\
&= \sum_{n=0}^{\infty} 2^n \mathbf{P}(X=n) = \sum_{n=0}^{\infty} 2^n \left(\frac{1}{2}\right)^{n+1} \\
&= \sum_{n=0}^{\infty} \frac{1}{2} = \infty.
\end{aligned}
$$

Hence, there is obviously something wrong with the classical theory of hazard games in this case.

Another important characteristic, besides $\mathbf{E}[X]$, of a random variable X, is the **variance Var$[X]$**, defined by

$$
\mathbf{Var}[X] = \mathbf{E}[(X-\mu)^2] \quad \text{where } \mu = \mathbf{E}[X]. \tag{2}
$$

The variance is, thus, the mean square deviation of X from its expectation. It can be computed either using the defining formula (2), or by the identity

$$
\mathbf{Var}[X] = \mathbf{E}[X^2] - (\mathbf{E}[X])^2 \tag{3}
$$

known as **Steiner's formula**.

There are various linear-like rules for working with expectations and variances. For expectations, we have

$$
\mathbf{E}[X_1 + \cdots + X_n] = \mathbf{E}[X_1] + \cdots + \mathbf{E}[X_n] \tag{4}
$$

and, if c is a constant,

$$
\mathbf{E}[cX] = c\mathbf{E}[X]. \tag{5}
$$

For variances, we have

$$
\mathbf{Var}[cX] = c^2 \mathbf{Var}[X] \tag{6}
$$

and, *when $X_1, \ldots, X_n$ are independent,*[1]

$$
\mathbf{Var}[X_1 + \cdots + X_n] = \mathbf{Var}[X_1] + \cdots + \mathbf{Var}[X_n]. \tag{7}
$$

Let us compute expectations and variances in some simple cases.

Example 1.2 Fix $p \in [0, 1]$, and let

$$
X = \begin{cases} 1 & \text{with probability } p \\ 0 & \text{with probability } 1-p. \end{cases}
$$

---

[1] Without this requirement, (7) *fails* in general.

Such an X is called a **Bernoulli (p) random variable**. The expectation of X becomes $\mathbf{E}[X] = 0 \cdot \mathbf{P}(X = 0) + 1 \cdot \mathbf{P}(X = 1) = p$. Furthermore, since X only takes the values 0 and 1, we have $X^2 = X$, so that $\mathbf{E}[X^2] = \mathbf{E}[X]$, and

$$\begin{aligned} \mathbf{Var}[X] &= \mathbf{E}[X^2] - (\mathbf{E}[X])^2 \\ &= p - p^2 = p(1 - p) \end{aligned}$$

using Steiner's formula (3).

Example 1.3 Let Y be the sum of n independent Bernoulli (p) random variables $X_1, \ldots, X_n$. (For instance, Y may be the number of heads in n tosses of a coin with heads-probability p.) Such a Y is said to be a **binomial (n, p) random variable**. Then, using (4) and (7), we get

$$\mathbf{E}[Y] = \mathbf{E}[X_1] + \cdots + \mathbf{E}[X_n] = np$$

and

$$\mathbf{Var}[Y] = \mathbf{Var}[X_1] + \cdots + \mathbf{Var}[X_n] = np(1 - p) \, .$$

Variances are useful, e.g., for bounding the probability that a random variable deviates by a large amount from its mean. We have, for instance, the following well-known result.

Theorem 1.1 (Chebyshev's inequality) *Let X be a random variable with mean μ and variance σ^2. For any $a > 0$, we have that the probability $\mathbf{P}[|X - \mu| \geq a]$ of a deviation from the mean of at least a, satisfies*

$$\mathbf{P}(|X - \mu| \geq a) \leq \frac{\sigma^2}{a^2} \, .$$

Proof Define another random variable Y by setting

$$Y = \begin{cases} a^2 & \text{if } |X - \mu| \geq a \\ 0 & \text{otherwise.} \end{cases}$$

Then we always have $Y \leq (X - \mu)^2$, so that $\mathbf{E}[Y] \leq \mathbf{E}[(X - \mu)^2]$. Furthermore, $\mathbf{E}[Y] = a^2 \mathbf{P}(|X - \mu| \geq a)$, so that

$$\begin{aligned} \mathbf{P}(|X - \mu| \geq a) &= \frac{\mathbf{E}[Y]}{a^2} \\ &\leq \frac{\mathbf{E}[(X - \mu)^2]}{a^2} \\ &= \frac{\mathbf{Var}[X]}{a^2} = \frac{\sigma^2}{a^2} \, . \end{aligned}$$

$\square$

Chebyshev's inequality will be used to prove a key result in Chapter 9 (Lemma 9.3). A more famous application of Chebyshev's inequality is in the proof of the following very famous and important result.

Theorem 1.2 (The Law of Large Numbers) *Let $X_1, X_2, \ldots$ be i.i.d. random variables with finite mean μ and finite variance σ^2. Let M_n denote the average of the first n X_i's, i.e., $M_n = \frac{1}{n}(X_1 + \cdots + X_n)$. Then, for any $\varepsilon > 0$, we have*

$$\lim_{n \to \infty} \mathbf{P}(|M_n - \mu| \geq \varepsilon) = 0 .$$

Proof Using (4) and (5) we get

$$\mathbf{E}[M_n] = \frac{1}{n}(\mu + \cdots + \mu) = \mu .$$

Similarly, (6) and (7) apply to show that

$$\mathbf{Var}[M_n] = \frac{1}{n^2}(\sigma^2 + \cdots + \sigma^2) = \frac{\sigma^2}{n} .$$

Hence, Chebyshev's inequality gives

$$\mathbf{P}(|M_n - \mu| \geq \varepsilon) \leq \frac{\sigma^2}{n\varepsilon^2}$$

which tends to 0 as $n \to \infty$. ☐

2

Markov chains

Let us begin with a simple example. We consider a "random walker" in a very small town consisting of four streets, and four street-corners v_1, v_2, v_3 and v_4 arranged as in Figure 1. At time 0, the random walker stands in corner v_1. At time 1, he flips a fair coin and moves immediately to v_2 or v_4 according to whether the coin comes up heads or tails. At time 2, he flips the coin again to decide which of the two adjacent corners to move to, with the decision rule that if the coin comes up heads, then he moves one step clockwise in Figure 1, while if it comes up tails, he moves one step counterclockwise. This procedure is then iterated at times 3, 4, ...

For each n, let X_n denote the index of the street-corner at which the walker stands at time n. Hence, $(X_0, X_1, \ldots)$ is a random process taking values in $\{1, 2, 3, 4\}$. Since the walker starts at time 0 in v_1, we have

$$\mathbf{P}(X_0 = 1) = 1 . \tag{8}$$

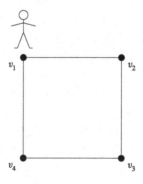

Fig. 1. A random walker in a very small town.

8

Next, he will move to v_2 or v_4 with probability $\frac{1}{2}$ each, so that

$$\mathbf{P}(X_1 = 2) = \frac{1}{2} \tag{9}$$

and

$$\mathbf{P}(X_1 = 4) = \frac{1}{2}. \tag{10}$$

To compute the distribution of X_n for $n \geq 2$ requires a little more thought; you will be asked to do this in Problem 2.1 below. To this end, it is useful to consider conditional probabilities. Suppose that at time n, the walker stands at, say, v_2. Then we get the conditional probabilities

$$\mathbf{P}(X_{n+1} = v_1 \mid X_n = v_2) = \frac{1}{2}$$

and

$$\mathbf{P}(X_{n+1} = v_3 \mid X_n = v_2) = \frac{1}{2},$$

because of the coin-flipping mechanism for deciding where to go next. In fact, we get the same conditional probabilities if we condition further on the full history of the process up to time n, i.e.,

$$\mathbf{P}(X_{n+1} = v_1 \mid X_0 = i_0, X_1 = i_1, \ldots, X_{n-1} = i_{n-1}, X_n = v_2) = \frac{1}{2}$$

and

$$\mathbf{P}(X_{n+1} = v_3 \mid X_0 = i_0, X_1 = i_1, \ldots, X_{n-1} = i_{n-1}, X_n = v_2) = \frac{1}{2}$$

for any choice of $i_0, \ldots, i_{n-1}$. (This is because the coin flip at time $n + 1$ is independent of all previous coin flips, and hence also independent of $X_0, \ldots, X_n$.) This phenomenon is called the **memoryless property**, also known as the **Markov property**: the conditional distribution of X_{n+1} given $(X_0, \ldots, X_n)$ depends only on X_n. Or in other words: to make the best possible prediction of what happens "tomorrow" (time $n + 1$), we only need to consider what happens "today" (time n), as the "past" (times $0, \ldots, n - 1$) gives no additional useful information.[2]

Another interesting feature of this random process is that the conditional distribution of X_{n+1} given that $X_n = v_2$ (say) is the same for all n. (This is because the mechanism that the walker uses to decide where to go next is the

---

[2] Please note that this is just a property of this particular mathematical model. It is *not* intended as general advice that we should "never worry about the past". Of course, we have every reason, in daily life as well as in politics, to try to learn as much as we can from history in order to make better decisions for the future!

same at all times.) This property is known as **time homogeneity**, or simply **homogeneity**.

These observations call for a general definition:

Definition 2.1 *Let P be a $k \times k$ matrix with elements $\{P_{i,j} : i, j = 1, \ldots, k\}$. A random process $(X_0, X_1, \ldots)$ with finite state space $S = \{s_1, \ldots, s_k\}$ is said to be a (**homogeneous**) **Markov chain with transition matrix** P, if for all n, all $i, j \in \{1, \ldots, k\}$ and all $i_0, \ldots, i_{n-1} \in \{1, \ldots, k\}$ we have*

$$\mathbf{P}(X_{n+1} = s_j \mid X_0 = s_{i_0}, X_1 = s_{i_1}, \ldots, X_{n-1} = s_{i_{n-1}}, X_n = s_i)$$
$$= \mathbf{P}(X_{n+1} = s_j \mid X_n = s_i)$$
$$= P_{i,j}.$$

The elements of the transition matrix P are called transition probabilities. The transition probability $P_{i,j}$ is the conditional probability of being in state s_j "tomorrow" given that we are in state s_i "today". The term "homogeneous" is often dropped, and taken for granted when talking about "Markov chains".

For instance, the random walk example above is a Markov chain, with state space $\{1, \ldots, 4\}$ and transition matrix

$$P = \begin{bmatrix} 0 & \frac{1}{2} & 0 & \frac{1}{2} \\ \frac{1}{2} & 0 & \frac{1}{2} & 0 \\ 0 & \frac{1}{2} & 0 & \frac{1}{2} \\ \frac{1}{2} & 0 & \frac{1}{2} & 0 \end{bmatrix}. \tag{11}$$

Every transition matrix satisfies

$$P_{i,j} \geq 0 \text{ for all } i, j \in \{1, \ldots, k\}, \tag{12}$$

and

$$\sum_{j=1}^{k} P_{i,j} = 1 \text{ for all } i \in \{1, \ldots, k\}. \tag{13}$$

Property (12) is just the fact that conditional probabilities are always nonnegative, and property (13) is that they sum to 1, i.e.,

$$\mathbf{P}(X_{n+1} = s_1 \mid X_n = s_i) + \mathbf{P}(X_{n+1} = s_2 \mid X_n = s_i) + \cdots$$
$$+ \mathbf{P}(X_{n+1} = s_k \mid X_n = s_i) = 1.$$

We next consider another important characteristic (besides the transition matrix) of a Markov chain $(X_0, X_1, \ldots)$, namely the **initial distribution**, which tells us how the Markov chain starts. The initial distribution is represented as a row vector $\mu^{(0)}$ given by

$$
\begin{aligned}
\mu^{(0)} &= (\mu_1^{(0)}, \mu_2^{(0)}, \dots, \mu_k^{(0)}) \\
&= (\mathbf{P}(X_0 = s_1), \mathbf{P}(X_0 = s_2), \dots, \mathbf{P}(X_0 = s_k)).
\end{aligned}
$$

Since $\mu^{(0)}$ represents a probability distribution, we have

$$
\sum_{i=1}^{k} \mu_i^{(0)} = 1.
$$

In the random walk example above, we have

$$
\mu^{(0)} = (1, 0, 0, 0) \tag{14}
$$

because of (8).

Similarly, we let the row vectors $\mu^{(1)}, \mu^{(2)}, \dots$ denote the distributions of the Markov chain at times $1, 2, \dots$, so that

$$
\begin{aligned}
\mu^{(n)} &= (\mu_1^{(n)}, \mu_2^{(n)}, \dots, \mu_k^{(n)}) \\
&= (\mathbf{P}(X_n = s_1), \mathbf{P}(X_n = s_2), \dots, \mathbf{P}(X_n = s_k)).
\end{aligned}
$$

For the random walk example, equations (9) and (10) tell us that

$$
\mu^{(1)} = (0, \tfrac{1}{2}, 0, \tfrac{1}{2}).
$$

It turns out that once we know the initial distribution $\mu^{(0)}$ and the transition matrix P, we can compute all the distributions $\mu^{(1)}, \mu^{(2)}, \dots$ of the Markov chain. The following result tells us that this is simply a matter of matrix multiplication. We write P^n for the nth power of the matrix P.

Theorem 2.1 *For a Markov chain* $(X_0, X_1, \dots)$ *with state space* $\{s_1, \dots, s_k\}$, *initial distribution* $\mu^{(0)}$ *and transition matrix* P, *we have for any* n *that the distribution* $\mu^{(n)}$ *at time* n *satisfies*

$$
\mu^{(n)} = \mu^{(0)} P^n. \tag{15}
$$

Proof Consider first the case $n = 1$. We get, for $j = 1, \dots, k$, that

$$
\begin{aligned}
\mu_j^{(1)} &= \mathbf{P}(X_1 = s_j) = \sum_{i=1}^{k} \mathbf{P}(X_0 = s_i, X_1 = s_j) \\
&= \sum_{i=1}^{k} \mathbf{P}(X_0 = s_i) \mathbf{P}(X_1 = s_j \mid X_0 = s_i) \\
&= \sum_{i=1}^{k} \mu_i^{(0)} P_{i,j} = (\mu^{(0)} P)_j
\end{aligned}
$$

where $(\mu^{(0)} P)_j$ denotes the jth element of the row vector $\mu^{(0)} P$. Hence $\mu^{(1)} = \mu^{(0)} P$.

To prove (15) for the general case, we use induction. Fix m, and suppose that (15) holds for $n = m$. For $n = m + 1$, we get

$$\mu_j^{(m+1)} = \mathbf{P}(X_{m+1} = s_j) = \sum_{i=1}^{k} \mathbf{P}(X_m = s_i, X_{m+1} = s_j)$$

$$= \sum_{i=1}^{k} \mathbf{P}(X_m = s_i)\mathbf{P}(X_{m+1} = s_j \mid X_m = s_i)$$

$$= \sum_{i=1}^{k} \mu_i^{(m)} P_{i,j} = (\mu^{(m)} P)_j$$

so that $\mu^{(m+1)} = \mu^{(m)} P$. But $\mu^{(m)} = \mu^{(0)} P^m$ by the induction hypothesis, so that

$$\mu^{(m+1)} = \mu^{(m)} P = \mu^{(0)} P^m P = \mu^{(0)} P^{m+1}$$

and the proof is complete.   □

Let us consider some more examples – two small ones, and one huge:

Example 2.1: The Gothenburg weather. It is sometimes claimed that the best way to predict tomorrow's weather[3] is simply to guess that it will be the same tomorrow as it is today. If we assume that this claim is correct,[4] then it is natural to model the weather as a Markov chain. For simplicity, we assume that there are only two kinds of weather: rain and sunshine. If the above predictor is correct 75% of the time (regardless of whether today's weather is rain or sunshine), then the weather forms a Markov chain with state space $S = \{s_1, s_2\}$ (with $s_1 =$ "rain" and $s_2 =$ "sunshine") and transition matrix

$$P = \begin{bmatrix} 0.75 & 0.25 \\ 0.25 & 0.75 \end{bmatrix}.$$

Example 2.2: The Los Angeles weather. Note that in Example 2.1, there is a perfect symmetry between "rain" and "sunshine", in the sense that the probability that today's weather will persist tomorrow is the same regardless of today's weather. This may be reasonably realistic in Gothenburg, but not in Los Angeles where sunshine is much more common than rain. A more reasonable transition matrix for the Los Angeles weather might therefore be (still with $s_1 =$ "rain" and $s_2 =$ "sunshine")

$$P = \begin{bmatrix} 0.5 & 0.5 \\ 0.1 & 0.9 \end{bmatrix}. \tag{16}$$

---

[3] Better than watching the weather forecast on TV.
[4] I doubt it.

Example 2.3: The Internet as a Markov chain. Imagine that you are surfing on the Internet, and that each time that you encounter a web page, you click on one of its hyperlinks chosen at random (uniformly). If X_n denotes where you are after n clicks, then $(X_0, X_1, \ldots)$ may be described as a Markov chain with state space S equal to the set of all web pages on the Internet, and transition matrix P given by

$$P_{ij} = \begin{cases} \frac{1}{d_i} & \text{if page } s_i \text{ has a link to page } s_j \\ 0 & \text{otherwise,} \end{cases}$$

where d_i is the number of links from page s_i. (To make this chain well-defined, we also need to define what happens if there are no links at all from s_i. We may, for instance, set $P_{ii} = 1$ (and $P_{ij} = 0$ for all $i \neq j$) in that case, meaning that when you encounter a page with no links, you are stuck.) This is of course a very complicated Markov chain (especially compared to Examples 2.1 and 2.2), but it has nevertheless turned out to be a useful model which under various simplifying assumptions admits interesting analysis.[5]

A recent variant (see Fagin *et al.* [Fa]) of this model is to take into account also the possibility to use "back buttons" in web browsers. However, the resulting process $(X_0, X_1, \ldots)$ is then no longer a Markov chain, since what happens when the back button is pressed depends not only on the present state X_n, but in general also on $X_0, \ldots, X_{n-1}$. Nevertheless, it turns out that this variant can be studied by a number of techniques from the theory of Markov chains. We will not say anything more about this model here.

A useful way to picture a Markov chain is its so-called **transition graph**. The transition graph consists of nodes representing the states of the Markov chain, and arrows between the nodes, representing transition probabilities. This is most easily explained by just showing the transition graphs of the examples considered so far. See Figure 2.

In all examples above, as well as in Definition 2.1, the "rule" for obtaining X_{n+1} from X_n did not change with time. In some situations, it is more realistic, or for other reasons more desirable,[6] to let this rule change with time. This brings us to the topic of **inhomogeneous Markov chains**, and the following definition, which generalizes Definition 2.1.

Definition 2.2 *Let $P^{(1)}, P^{(2)}, \ldots$ be a sequence of $k \times k$ matrices, each of which satisfies (12) and (13). A random process $(X_0, X_1, \ldots)$ with finite state space $S = \{s_1, \ldots, s_k\}$ is said to be an* **inhomogeneous Markov chain with transition matrices** $P^{(1)}, P^{(2)}, \ldots$, *if for all n, all $i, j \in \{1, \ldots, k\}$ and all*

---

[5] It may also seem like a very big Markov chain. However, the devoted reader will soon know how to carry out (not just in principle, but also in practice) computer simulations of much bigger Markov chains – see, e.g., Problem 7.2.

[6] Such as in the simulated annealing algorithms of Chapter 13.

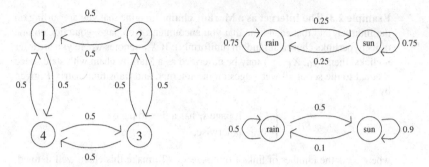

Fig. 2. Transition graphs for the random walker in Figure 1, and for Examples 2.1 and 2.2.

$i_0, \ldots, i_{n-1} \in \{1, \ldots, k\}$ *we have*

$$\mathbf{P}(X_{n+1} = s_j \mid X_0 = s_{i_0}, X_1 = s_{i_1}, \ldots, X_{n-1} = s_{i_{n-1}}, X_n = s_i)$$
$$= \mathbf{P}(X_{n+1} = s_j \mid X_n = s_i)$$
$$= P_{i,j}^{(n+1)}.$$

Example 2.4: A refined model for the Gothenburg weather. There are of course many ways in which the crude model in Example 2.1 can be made more realistic. One way is to take into account seasonal changes: it does not seem reasonable to disregard whether the calendar says "January" or "July" when predicting tomorrow's weather. To this end, we extend the state space to $\{s_1, s_2, s_3\}$, where $s_1 = $ "rain" and $s_2 = $ "sunshine" as before, and $s_3 = $ "snow". Let

$$P_{summer} = \begin{bmatrix} 0.75 & 0.25 & 0 \\ 0.25 & 0.75 & 0 \\ 0.5 & 0.5 & 0 \end{bmatrix} \text{ and } P_{winter} = \begin{bmatrix} 0.5 & 0.3 & 0.2 \\ 0.15 & 0.7 & 0.15 \\ 0.2 & 0.3 & 0.5 \end{bmatrix},$$

and assume that the weather evolves according to P_{summer} in May–September, and according to P_{winter} in October–April. This is an inhomogeneous Markov chain model for the Gothenburg weather. Note that in May–September, the model behaves exactly like the one in Example 2.1, except for some possible residual snowy weather on May 1.

The following result, which is a generalization of Theorem 2.1, tells us how to compute the distributions $\mu^{(1)}, \mu^{(2)}, \ldots$ at times $1, 2, \ldots$ of an inhomogeneous Markov chain with initial distribution $\mu^{(0)}$ and transition matrices $P^{(1)}, P^{(2)}, \ldots$.

Theorem 2.2 *Suppose that* $(X_0, X_1, \ldots)$ *is an inhomogeneous Markov chain with state space* $\{s_1, \ldots, s_k\}$, *initial distribution* $\mu^{(0)}$ *and transition matrices*

$P^{(1)}, P^{(2)}, \ldots$ *For any n, we then have that*

$$\mu^{(n)} = \mu^{(0)} P^{(1)} P^{(2)} \cdots P^{(n)}.$$

Proof Follows by a similar calculation as in the proof of Theorem 2.1. $\quad\square$

Problems

2.1 (5) Consider the Markov chain corresponding to the random walker in Figure 1, with transition matrix P and initial distribution $\mu^{(0)}$ given by (11) and (14).

(a) Compute the square P^2 of the transition matrix P. How can we interpret P^2? (See Theorem 2.1, or glance ahead at Problem 2.5.)

(b) Prove by induction that

$$\mu^{(n)} = \begin{cases} (0, \frac{1}{2}, 0, \frac{1}{2}) & \text{for } n = 1, 3, 5, \ldots \\ (\frac{1}{2}, 0, \frac{1}{2}, 0) & \text{for } n = 2, 4, 6, \ldots. \end{cases}$$

2.2 (2) Suppose that we modify the random walk example in Figure 1 as follows. At each integer time, the random walker tosses *two* coins. The first coin is to decide whether to stay or go. If it comes up heads, he stays where he is, whereas if it comes up tails, he lets the second coin decide whether he should move one step clockwise, or one step counterclockwise. Write down the transition matrix, and draw the transition graph, for this new Markov chain.

2.3 (5) Consider Example 2.1 (the Gothenburg weather), and suppose that the Markov chain starts on a rainy day, so that $\mu^{(0)} = (1, 0)$.

(a) Prove by induction that

$$\mu^{(n)} = (\tfrac{1}{2}(1 + 2^{-n}), \tfrac{1}{2}(1 - 2^{-n}))$$

for every n.

(b) What happens to $\mu^{(n)}$ in the limit as n tends to infinity?

2.4 (6)

(a) Consider Example 2.2 (the Los Angeles weather), and suppose that the Markov chain starts with initial distribution $(\frac{1}{6}, \frac{5}{6})$. Show that $\mu^{(n)} = \mu^{(0)}$ for any n, so that in other words the distribution remains the same at all times.[7]

(b) Can you find an initial distribution for the Markov chain in Example 2.1 for which we get similar behavior as in (a)? Compare this result to the one in Problem 2.3 (b).

2.5 (6) Let $(X_0, X_1, \ldots)$ be a Markov chain with state space $\{s_1, \ldots, s_k\}$ and transition matrix P. Show, by arguing as in the proof of Theorem 2.1, that for any $m, n \geq 0$ we have

$$\mathbf{P}(X_{m+n} = s_j \mid X_m = s_i) = (P^n)_{i,j}.$$

---

[7] Such a Markov chain is said to be in **equilibrium**, and its distribution is said to be **stationary**. This is a very important topic, which will be treated carefully in Chapter 5.

2.6 (8) Functions of Markov chains are not always Markov chains. Let $(X_0, X_1, \ldots)$ be a Markov chain with state space $\{s_1, s_2, s_3\}$, transition matrix

$$P = \begin{bmatrix} 0 & 1 & 0 \\ 0 & 0 & 1 \\ 1 & 0 & 0 \end{bmatrix}$$

and initial distribution $\mu^{(0)} = (\frac{1}{3}, \frac{1}{3}, \frac{1}{3})$. For each n, define

$$Y_n = \begin{cases} 0 & \text{if } X_n = s_1 \\ 1 & \text{otherwise.} \end{cases}$$

Show that $(Y_0, Y_1, \ldots)$ is *not* a Markov chain.

2.7 (9) Markov chains sampled at regular intervals are Markov chains. Let $(X_0, X_1, \ldots)$ be a Markov chain with transition matrix P.

(a) Define $(Y_0, Y_1, \ldots)$ by setting $Y_n = X_{2n}$ for each n. Show that $(Y_0, Y_1, \ldots)$ is a Markov chain with transition matrix P^2.

(b) Find an appropriate generalization of the result in (a) to the situation where we sample every k^{th} (rather than every second) value of $(X_0, X_1, \ldots)$.

3

Computer simulation of Markov chains

A key matter in many (most?) practical applications of Markov theory is the ability to simulate Markov chains on a computer. This chapter deals with how that can be done.

We begin by stating a lie:

> In most high-level programming languages, we have access to some random number generator producing a sequence $U_0, U_1, \ldots$ of i.i.d. random variables, uniformly distributed on the unit interval $[0, 1]$.

This is a lie for at least two reasons:

(A) The numbers $U_0, U_1, \ldots$ obtained from random number generators are not uniformly distributed on $[0, 1]$. Typically, they have a finite binary (or decimal) expansion, and are therefore rational. In contrast, it can be shown that a random variable which (truly) is uniformly distributed on $[0, 1]$ (or in fact any continuous random variable) is irrational with probability 1.

(B) $U_0, U_1, \ldots$ are not even random! Rather, they are obtained by some deterministic procedure. For this reason, random number generators are sometimes (and more accurately) called pseudo-random number generators.[8]

The most important of these objections is (B), because (A) tends not to be a very big problem when the number of binary or decimal digits is reasonably large (say, 32 bits). Over the decades, a lot of effort has been put into constructing (pseudo-)random number generators whose output is as indistinguishable

---

[8] There are also various physically generated sequences of random-looking numbers (see, e.g., the web sites http://lavarand.sgi.com/ and http://www.fourmilab.ch/hotbits/) that may be used instead of the usual pseudo-random number generators. I recommend, however, a healthy dose of skepticism towards claims that these sequences are in some sense "truly" random.

as possible from a true i.i.d. sequence of uniform $[0, 1]$ random variables. Today, there exist generators which appear to do this very well (passing all of a number of standard statistical tests for such generators), and for this reason, *we shall simply make the (incorrect) assumption that we have access to an i.i.d. sequence of uniform $[0, 1]$ random variables $U_0, U_1, \ldots$.* Although we shall not deal any further with the pseudo-randomness issue in the remainder of these notes (except for providing a couple of relevant references in the final chapter), we should always keep in mind that it is a potential source of errors in computer simulation.[9]

Let us move on to the core topic of this chapter: How do we simulate a Markov chain $(X_0, X_1, \ldots)$ with given state space $S = \{s_1, \ldots, s_k\}$, initial distribution $\mu^{(0)}$ and transition matrix P? As the reader probably has guessed by now, the random numbers $U_0, U_1, \ldots$ form a main ingredient. The other main ingredients are two functions, which we call the **initiation function** and the **update function**.

The initiation function $\psi : [0, 1] \to S$ is a function from the unit interval to the state space S, which we use to generate the starting value X_0. We assume

(i) that ψ is piecewise constant (i.e., that $[0, 1]$ can be split into finitely many subintervals in such a way that ψ is constant on each interval), and

(ii) that for each $s \in S$, the total length of the intervals on which $\psi(x) = s$ equals $\mu^{(0)}(s)$.

Another way to state property (ii) is that

$$\int_0^1 \mathbf{I}_{\{\psi(x)=s\}} \, dx = \mu^{(0)}(s) \tag{17}$$

for each $s \in S$; here $\mathbf{I}_{\{\psi(x)=s\}}$ is the so-called **indicator function** of $\{\psi(x) = s\}$, meaning that

$$\mathbf{I}_{\{\psi(x)=s\}} = \begin{cases} 1 & \text{if } \psi(x) = s \\ 0 & \text{otherwise.} \end{cases}$$

Provided that we have such a function ψ, we can generate X_0 from the first random number U_0 by setting $X_0 = \psi(U_0)$. This gives the correct distribution of X_0, because for any $s \in S$ we get

$$\mathbf{P}(X_0 = s) = \mathbf{P}(\psi(U_0) = s) = \int_0^1 \mathbf{I}_{\{\psi(x)=s\}} \, dx = \mu^{(0)}(s)$$

---

[9] A misunderstanding that I have encountered more than once is that a pseudo-random number generator is good if its period (the time until it repeats itself) is long, i.e., longer than the number of random numbers needed in a particular application. But this is far from sufficient, and many other things can go wrong. For instance, certain patterns may occur too frequently (or all the time).

using (17). Hence, we call ψ a **valid** initiation function for the Markov chain $(X_0, X_1, \ldots)$ if (17) holds for all $s \in S$.

Valid initiation functions are easy to construct: With $S = \{s_1, \ldots, s_k\}$ and initial distribution $\mu^{(0)}$, we can set

$$
\psi(x) = \begin{cases}
s_1 & \text{for } x \in [0, \mu^{(0)}(s_1)) \\
s_2 & \text{for } x \in [\mu^{(0)}(s_1), \mu^{(0)}(s_1) + \mu^{(0)}(s_2)) \\
\vdots & \vdots \\
s_i & \text{for } x \in \left[\sum_{j=1}^{i-1} \mu^{(0)}(s_j), \sum_{j=1}^{i} \mu^{(0)}(s_j) \right) \\
\vdots & \vdots \\
s_k & \text{for } x \in \left[\sum_{j=1}^{k-1} \mu^{(0)}(s_j), 1 \right].
\end{cases} \tag{18}
$$

We need to verify that this choice of ψ satisfies properties (i) and (ii) above. Property (i) is obvious. As to property (ii), it suffices to check that (17) holds. It does hold, since

$$
\int_0^1 \mathbf{I}_{\{\psi(x)=s_i\}} \, dx = \sum_{j=1}^{i} \mu^{(0)}(s_j) - \sum_{j=1}^{i-1} \mu^{(0)}(s_j) = \mu^{(0)}(s_i)
$$

for $i = 1, \ldots, k$. This means that ψ as defined in (18) is a valid initiation function for the Markov chain $(X_0, X_1, \ldots)$.

So now we know how to generate the starting value X_0. If we also figure out how to generate X_{n+1} from X_n for any n, then we can use this procedure iteratively to get the whole chain $(X_0, X_1, \ldots)$. To get from X_n to X_{n+1}, we use the random number U_{n+1} and an **update function** $\phi : S \times [0, 1] \to S$, which takes as input a state $s \in S$ and a number between 0 and 1, and produces another state $s' \in S$ as output. Similarly as for the initiation function ψ, we need ϕ to obey certain properties, namely

(i) that for fixed s_i, the function $\phi(s_i, x)$ is piecewise constant (when viewed as a function of x), and

(ii) that for each fixed $s_i, s_j \in S$, the total length of the intervals on which $\phi(s_i, x) = s_j$ equals $P_{i,j}$.

Again, as for the initiation function, property (ii) can be rewritten as

$$
\int_0^1 \mathbf{I}_{\{\phi(s_i, x)=s_j\}} \, dx = P_{i,j} \tag{19}
$$

for all $s_i, s_j \in S$. If the update function ϕ satisfies (19), then

$$
\begin{aligned}
\mathbf{P}(X_{n+1} = s_j \mid X_n = s_i) &= \mathbf{P}(\phi(s_i, U_{n+1}) = s_j \mid X_n = s_i) \quad (20) \\
&= \mathbf{P}(\phi(s_i, U_{n+1}) = s_j) \\
&= \int_0^1 \mathbf{I}_{\{\phi(s_i, x) = s_j\}}\, dx = P_{i,j}\,.
\end{aligned}
$$

The reason that the conditioning in (20) can be dropped is that U_{n+1} is independent of $(U_0, \ldots, U_n)$, and hence also of X_n. The same argument shows that the conditional probability remains the same if we condition further on the values $(X_0, X_1, \ldots, X_{n-1})$. Hence, this gives a correct simulation of the Markov chain. A function ϕ satisfying (19) is therefore said to be a valid update function for the Markov chain $(X_0, X_1, \ldots)$.

It remains to construct such a valid update function, but this is no harder than the construction of a valid initiation function: Set, for each $s_i \in S$,

$$
\phi(s_i, x) = \begin{cases}
s_1 & \text{for } x \in [0, P_{i,1}) \\
s_2 & \text{for } x \in [P_{i,1},\ P_{i,1} + P_{i,2}) \\
\ \vdots & \quad \vdots \\
s_j & \text{for } x \in \left[\sum_{l=1}^{j-1} P_{i,l},\ \sum_{l=1}^{j} P_{i,l} \right) \\
\ \vdots & \quad \vdots \\
s_k & \text{for } x \in \left[\sum_{l=1}^{k-1} P_{i,l},\ 1 \right].
\end{cases}
\quad (21)
$$

To see that this is a valid update function, note that for any $s_i, s_j \in S$, we have

$$
\int_0^1 \mathbf{I}_{\{\phi(s_i, x) = s_j\}}\, dx = \sum_{l=1}^{j} P_{i,l} - \sum_{l=1}^{j-1} P_{i,l} = P_{i,j}\,.
$$

Thus, we have a complete recipe for simulating a Markov chain: First construct valid initiation and update functions ψ and ϕ (for instance as in (18) and (21)), and then set

$$
\begin{aligned}
X_0 &= \psi(U_0) \\
X_1 &= \phi(X_0, U_1) \\
X_2 &= \phi(X_1, U_2) \\
X_3 &= \phi(X_2, U_3)
\end{aligned}
$$

and so on.

Let us now see how the above works for a simple example.

Example 3.1: Simulating the Gothenburg weather. Consider the Markov chain in Example 2.1, whose state space is $S = \{s_1, s_2\}$ where $s_1 =$ "rain" and $s_2 =$

"sunshine", and whose transition matrix is given by

$$P = \begin{bmatrix} 0.75 & 0.25 \\ 0.25 & 0.75 \end{bmatrix}.$$

Suppose we start the Markov chain on a rainy day (as in Problem 2.3), so that $\mu^{(0)} = (1, 0)$. To simulate this Markov chain using the above scheme, we apply (18) and (21) to get the initiation function

$$\psi(x) = s_1 \quad \text{for all } x,$$

and update function given by

$$\phi(s_1, x) = \begin{cases} s_1 & \text{for } x \in [0, 0.75) \\ s_2 & \text{for } x \in [0.75, 1] \end{cases}$$

and

$$\phi(s_2, x) = \begin{cases} s_1 & \text{for } x \in [0, 0.25) \\ s_2 & \text{for } x \in [0.25, 1]. \end{cases} \tag{22}$$

Before closing this chapter, let us finally point out how the above method can be generalized to cope with simulation of inhomogeneous Markov chains. Let $(X_0, X_1, \ldots)$ be an inhomogeneous Markov chain with state space $S = \{s_1, \ldots, s_k\}$, initial distribution $\mu^{(0)}$, and transition matrices $P^{(0)}, P^{(1)}, \ldots$. We can then obtain the initiation function ψ and the starting value X_0 as in the homogeneous case. The updating is done similarly as in the homogeneous case, except that since the chain is inhomogeneous, we need several different updating functions $\phi^{(1)}, \phi^{(2)}, \ldots$, and for these we need to have

$$\int_0^1 \mathbf{I}_{\{\phi^{(n)}(s_i, x) = s_j\}}(x)\, dx = P_{i,j}^{(n)}$$

for each n and each $s_i, s_j \in S$. Such functions can be obtained by the obvious generalization of (21): Set

$$\phi^{(n)}(s_i, x) = \begin{cases} s_1 & \text{for } x \in [0, P_{i,1}^{(n)}) \\ s_2 & \text{for } x \in [P_{i,1}^{(n)}, P_{i,1}^{(n)} + P_{i,2}^{(n)}) \\ \vdots & \vdots \\ s_j & \text{for } x \in \left[\sum_{l=1}^{j-1} P_{i,l}^{(n)}, \sum_{l=1}^{j} P_{i,l}^{(n)} \right) \\ \vdots & \vdots \\ s_k & \text{for } x \in \left[\sum_{l=1}^{k-1} P_{i,l}^{(n)}, 1 \right]. \end{cases}$$

The inhomogeneous Markov chain is then simulated by setting

$$
\begin{aligned}
X_0 &= \psi(U_0) \\
X_1 &= \phi^{(1)}(X_0, U_1) \\
X_2 &= \phi^{(2)}(X_1, U_2) \\
X_3 &= \phi^{(3)}(X_2, U_3)
\end{aligned}
$$

and so on.

Problems

3.1 **(7*)**

(a) Find valid initiation and update functions for the Markov chain in Example 2.2 (the Los Angeles weather), with starting distribution $\mu^{(0)} = (\frac{1}{2}, \frac{1}{2})$.

(b) Write a computer program for simulating the Markov chain, using the initiation and update functions in (a).

(c) For $n \geq 1$, define Y_n to be the proportion of rainy days up to time n, i.e.,

$$
Y_n = \frac{1}{n+1} \sum_{i=0}^{n} \mathbf{I}_{\{X_i = s_1\}}.
$$

Simulate the Markov chain for (say) 1000 steps, and plot how Y_n evolves with time. What seems to happen to Y_n when n gets large? (Compare with Problem 2.4 (a).)

3.2 **(3)** **The choice of update function is not necessarily unique.** Consider Example 3.1 (simulating the Gothenburg weather). Show that we get another valid update function if we replace (22) by

$$
\phi(s_2, x) = \begin{cases} s_2 & \text{for } x \in [0, 0.75) \\ s_1 & \text{for } x \in [0.75, 1]. \end{cases}
$$

4

Irreducible and aperiodic Markov chains

For several of the most interesting results in Markov theory, we need to put certain assumptions on the Markov chains we are considering. It is an important task, in Markov theory just as in all other branches of mathematics, to find conditions that on the one hand are strong enough to have useful consequences, but on the other hand are weak enough to hold (and be easy to check) for many interesting examples. In this chapter, we will discuss two such conditions on Markov chains: **irreducibility** and **aperiodicity**. These conditions are of central importance in Markov theory, and in particular they play a key role in the study of stationary distributions, which is the topic of Chapter 5. We shall, for simplicity, discuss these notions in the setting of homogeneous Markov chains, although they do have natural extensions to the more general setting of inhomogeneous Markov chains.

We begin with irreducibility, which, loosely speaking, is the property that "all states of the Markov chain can be reached from all others". To make this more precise, consider a Markov chain $(X_0, X_1, \ldots)$ with state space $S = \{s_1, \ldots, s_k\}$ and transition matrix P. We say that a state s_i **communicates** with another state s_j, writing $s_i \rightarrow s_j$, if the chain has positive probability[10] of ever reaching s_j when we start from s_i. In other words, s_i communicates with s_j if there exists an n such that

$$\mathbf{P}(X_{m+n} = s_j \mid X_m = s_i) > 0\,.$$

By Problem 2.5, this probability is independent of m (due to the homogeneity of the Markov chain), and equals $(P^n)_{i,j}$.

If $s_i \rightarrow s_j$ and $s_j \rightarrow s_i$, then we say that the states s_i and s_j **intercommunicate**, and write $s_i \leftrightarrow s_j$. This takes us directly to the definition of irreducibility.

---

[10] Here and henceforth, by "positive probability", we always mean *strictly* positive probability.

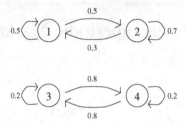

Fig. 3. Transition graph for the Markov chain in Example 4.1.

Definition 4.1 *A Markov chain* $(X_0, X_1, \ldots)$ *with state space* $S = \{s_1, \ldots, s_k\}$ *and transition matrix* P *is said to be* **irreducible** *if for all* $s_i, s_j \in S$ *we have that* $s_i \leftrightarrow s_j$. *Otherwise the chain is said to be* **reducible**.

Another way of phrasing the definition would be to say that the chain is irreducible if for any $s_i, s_j \in S$ we can find an n such that $(P^n)_{i,j} > 0$.

An easy way to verify that a Markov chain is irreducible is to look at its transition graph, and check that from each state there is a sequence of arrows leading to any other state. A glance at Figure 2 thus reveals that the Markov chains in Examples 2.1 and 2.2, as well as the random walk example in Figure 1, are all irreducible.[11] Let us next have a look at an example which is *not* irreducible:

    Example 4.1: A reducible Markov chain.   Consider a Markov chain $(X_0, X_1, \ldots)$ with state space $S = \{1, 2, 3, 4\}$ and transition matrix

$$P = \begin{bmatrix} 0.5 & 0.5 & 0 & 0 \\ 0.3 & 0.7 & 0 & 0 \\ 0 & 0 & 0.2 & 0.8 \\ 0 & 0 & 0.8 & 0.2 \end{bmatrix}.$$

By taking a look at its transition graph (see Figure 3), we immediately see that if the chain starts in state 1 or state 2, then it is restricted to states 1 and 2 forever. Similarly, if it starts in state 3 or state 4, then it can never leave the subset $\{3, 4\}$ of the state space. Hence, the chain is reducible.

    Note that if the chain starts in state 1 or state 2, then it behaves exactly as if it were a Markov chain with state space $\{1, 2\}$ and transition matrix

$$\begin{bmatrix} 0.5 & 0.5 \\ 0.3 & 0.7 \end{bmatrix}.$$

If it starts in state 3 or state 4, then it behaves like a Markov chain with state space

---

[11] Some care is still needed; see Problem 4.1.

{3, 4} and transition matrix

$$\begin{bmatrix} 0.2 & 0.8 \\ 0.8 & 0.2 \end{bmatrix}.$$

This illustrates a characteristic feature of reducible Markov chains, which also explains the term "reducible": If a Markov chain is reducible, then the analysis of its long-term behavior can be reduced to the analysis of the long-term behavior of one or more Markov chains with smaller state space.

We move on to consider the concept of aperiodicity. For a finite or infinite set $\{a_1, a_2, \ldots\}$ of positive integers, we write $\gcd\{a_1, a_2, \ldots\}$ for the greatest common divisor of $a_1, a_2, \ldots$. The **period** $d(s_i)$ of a state $s_i \in S$ is defined as

$$d(s_i) = \gcd\{n \geq 1 : (P^n)_{i,i} > 0\}.$$

In words, the period of s_i is the greatest common divisor of the set of times that the chain can return (i.e., has positive probability of returning) to s_i, given that we start with $X_0 = s_i$. If $d(s_i) = 1$, then we say that the state s_i is **aperiodic**.

Definition 4.2 *A Markov chain is said to be* **aperiodic** *if all its states are aperiodic. Otherwise the chain is said to be* **periodic**.

Consider for instance Example 2.1 (the Gothenburg weather). It is easy to check that regardless of whether the weather today is rain or sunshine, we have for any n that the probability of having the same weather n days later is strictly positive. Or, expressed more compactly: $(P^n)_{i,i} > 0$ for all n and all states s_i.[12] This obviously implies that the Markov chain in Example 2.1 is aperiodic. Of course, the same reasoning applies to Example 2.2 (the Los Angeles weather).

On the other hand, let us consider the random walk example in Figure 1, where the random walker stands in corner v_1 at time 0. Clearly, he has to take an even number of steps in order to get back to v_1. This means that $(P^n)_{1,1} > 0$ only for $n = 2, 4, 6, \ldots$. Hence,

$$\gcd\{n \geq 1 : (P^n)_{i,i} > 0\} = \gcd\{2, 4, 6, \ldots\} = 2,$$

and the chain is therefore periodic.

One reason for the usefulness of aperiodicity is the following result.

Theorem 4.1 *Suppose that we have an aperiodic Markov chain $(X_0, X_1, \ldots)$ with state space $S = \{s_1, \ldots, s_k\}$ and transition matrix P. Then there exists an $N < \infty$ such that*

$$(P^n)_{i,i} > 0$$

---

[12] By a variant of Problem 2.3 (a), we in fact have that $(P^n)_{i,i} = \frac{1}{2}(1 + 2^{-n})$.

for all $i \in \{1, \ldots, k\}$ and all $n \geq N$.

To prove this result, we shall borrow the following lemma from number theory.

Lemma 4.1 *Let $A = \{a_1, a_2, \ldots\}$ be a set of positive integers which is*

(i) *nonlattice, meaning that $\gcd\{a_1, a_2, \ldots\} = 1$, and*

(ii) *closed under addition, meaning that if $a \in A$ and $a' \in A$, then $a + a' \in A$.*

Then there exists an integer $N < \infty$ such that $n \in A$ for all $n \geq N$.

Proof See, e.g., the appendix of Brémaud [B].   □

Proof of Theorem 4.1 For $s_i \in S$, let $A_i = \{n \geq 1 : (P^n)_{i,i} > 0\}$, so that in other words A_i is the set of possible return times to state s_i starting from s_i. We assumed that the Markov chain is aperiodic, and therefore the state s_i is aperiodic, so that A_i is nonlattice. Furthermore, A_i is closed under addition, for the following reason: If $a, a' \in A_i$, then $\mathbf{P}(X_a = s_i \mid X_0 = s_i) > 0$ and $\mathbf{P}(X_{a+a'} = s_i \mid X_a = s_i) > 0$. This implies that

$$
\begin{aligned}
\mathbf{P}(X_{a+a'} = s_i \mid X_0 = s_i) &\geq \mathbf{P}(X_a = s_i, X_{a+a'} = s_i \mid X_0 = s_i) \\
&= \mathbf{P}(X_a = s_i \mid X_0 = s_i)\mathbf{P}(X_{a+a'} = s_i \mid X_a = s_i) \\
&> 0
\end{aligned}
$$

so that $a + a' \in A_i$.

In summary, A_i satisfies assumptions (i) and (ii) of Lemma 4.1, which therefore implies that there exists an integer $N_i < \infty$ such that $(P^n)_{i,i} > 0$ for all $n \geq N_i$.

Theorem 4.1 now follows with $N = \max\{N_1, \ldots, N_k\}$.   □

By combining aperiodicity and irreducibility, we get the following important result, which will be used in the next chapter to prove the so-called Markov chain convergence theorem (Theorem 5.2).

Corollary 4.1 *Let $(X_0, X_1, \ldots)$ be an irreducible and aperiodic Markov chain with state space $S = \{s_1, \ldots, s_k\}$ and transition matrix P. Then there exists an $M < \infty$ such that $(P^n)_{i,j} > 0$ for all $i, j \in \{1, \ldots, k\}$ and all $n \geq M$.*

Proof By the assumed aperiodicity and Theorem 4.1, there exists an integer $N < \infty$ such that $(P^n)_{i,i} > 0$ for all $i \in \{1, \ldots, k\}$ and all $n \geq N$. Fix two states $s_i, s_j \in S$. By the assumed irreducibility, we can find some $n_{i,j}$ such that $(P^{n_{i,j}})_{i,j} > 0$. Let $M_{i,j} = N + n_{i,j}$. For any $m \geq M_{i,j}$, we have

$$\mathbf{P}(X_m = s_j \mid X_0 = s_i) \geq \mathbf{P}(X_{m-n_{i,j}} = s_i, X_m = s_j \mid X_0 = s_i)$$
$$= \mathbf{P}(X_{m-n_{i,j}} = s_i \mid X_0 = s_i)\mathbf{P}(X_m = s_j \mid X_{m-n_{i,j}} = s_i)$$
$$> 0 \qquad\qquad (23)$$

(the first factor in the second line of (23) is positive because $m - n_{i,j} \geq N$, and the second is positive by the choice of $n_{i,j}$). Hence, we have shown that $(P^m)_{i,j} > 0$ for all $m \geq M_{i,j}$. The corollary now follows with

$$M = \max\{M_{1,1}, M_{1,2} \ldots, M_{1,k}, M_{2,1}, \ldots, M_{k,k}\}.$$

$\square$

Problems

4.1 (3) Consider the Markov chain $(X_0, X_1, \ldots)$ with state space $S = \{s_1, s_2\}$ and transition matrix

$$P = \begin{bmatrix} \frac{1}{2} & \frac{1}{2} \\ 0 & 1 \end{bmatrix}.$$

(a) Draw the transition graph of this Markov chain.
(b) Show that the Markov chain is *not* irreducible (even though the transition matrix looks in some sense connected).
(c) What happens to X_n in the limit as $n \to \infty$?

4.2 (3) Show that if a Markov chain is irreducible and has a state s_i such that $P_{ii} > 0$, then it is also aperiodic.

4.3 (4) **Random chess moves.**

(a) Consider a chessboard with a lone white king making random moves, meaning that at each move, he picks one of the possible squares to move to, uniformly at random. Is the corresponding Markov chain irreducible and/or aperiodic?
(b) Same question, but with the king replaced by a bishop.
(c) Same question, but instead with a knight.

4.4 (6) **Oriented random walk on a torus.** Let a and b be positive integers, and consider the Markov chain with state space

$$\{(x, y) : x \in \{0, \ldots, a - 1\}, y \in \{0, \ldots, b - 1\}\},$$

and the following transition mechanism: If the chain is in state (x, y) at time n, then at time $n + 1$ it moves to $((x + 1) \bmod a, y)$ or $(x, (y + 1) \bmod b)$ with probability $\frac{1}{2}$ each.

(a) Show that this Markov chain is irreducible.
(b) Show that it is aperiodic if and only if $\gcd(a, b) = 1$.

5

Stationary distributions

In this chapter, we consider one of the central issues in Markov theory: asymptotics for the long-term behavior of Markov chains. What can we say about a Markov chain that has been running for a long time? Can we find interesting limit theorems?

If $(X_0, X_1, \ldots)$ is any nontrivial Markov chain, then the value of X_n will keep fluctuating infinitely many times as $n \to \infty$, and therefore we cannot hope to get results about X_n converging to a limit.[13] However, we may hope that the *distribution* of X_n settles down to a limit. This is indeed the case if the Markov chain is irreducible and aperiodic, which is what the main result of this chapter, the so-called Markov chain convergence theorem (Theorem 5.2), says.

Let us for a moment go back to the Markov chain in Example 2.2 (the Los Angeles weather), with state space $\{s_1, s_2\}$ and transition matrix given by (16). We saw in Problem 2.4 (a) that if we let the initial distribution $\mu^{(0)}$ be given by $\mu^{(0)} = (\frac{1}{6}, \frac{5}{6})$, then this distribution is preserved for all times, i.e., $\mu^{(n)} = \mu^{(0)}$ for all n. By some experimentation, we can easily convince ourselves that no other choice of initial distribution $\mu^{(0)}$ for this chain has the same property (try it!). Apparently, the distribution $(\frac{1}{6}, \frac{5}{6})$ plays a special role for this Markov chain, and we call it a **stationary distribution**.[14] The general definition is as follows.

Definition 5.1 *Let* $(X_0, X_1, \ldots)$ *be a Markov chain with state space* $\{s_1, \ldots, s_k\}$ *and transition matrix* P. *A row vector* $\pi = (\pi_1, \ldots, \pi_k)$ *is said to be a* **stationary distribution** *for the Markov chain, if it satisfies*

---

[13] That is, unless there is some state s_i of the Markov chain with the property that $P_{ii} = 1$; recall Problem 4.1 (c).

[14] Another term which is used by many authors for the same thing is **invariant distribution**. Yet another term is **equilibrium distribution**.

(i) $\pi_i \geq 0$ *for* $i = 1, \ldots, k$, *and* $\sum_{i=1}^{k} \pi_i = 1$, *and*

(ii) $\pi P = \pi$, *meaning that* $\sum_{i=1}^{k} \pi_i P_{i,j} = \pi_j$ *for* $j = 1, \ldots, k$.

Property (i) simply means that π should describe a probability distribution on $\{s_1, \ldots, s_k\}$. Property (ii) implies that if the initial distribution $\mu^{(0)}$ equals π, then the distribution $\mu^{(1)}$ of the chain at time 1 satisfies

$$\mu^{(1)} = \mu^{(0)} P = \pi P = \pi \, ,$$

and by iterating we see that $\mu^{(n)} = \pi$ for every n.

Since the definition of a stationary distribution really only depends on the transition matrix P, we also sometimes say that a distribution π satisfying the assumptions (i) and (ii) in Definition 5.1 is **stationary for the matrix** P (rather than for the Markov chain).

The rest of this chapter will deal with three issues: the **existence** of stationary distributions, the **uniqueness** of stationary distributions, and the **convergence** to stationarity starting from any initial distribution. We shall work under the conditions introduced in the previous chapter (irreducibility and aperiodicity), although for some of the results these conditions can be relaxed somewhat.[15] We begin with the existence issue.

Theorem 5.1 (Existence of stationary distributions) *For any irreducible and aperiodic Markov chain, there exists at least one stationary distribution.*

To prove this existence theorem, we first need to prove a lemma concerning **hitting times** for Markov chains. If a Markov chain $(X_0, X_1, \ldots)$ with state space $\{s_1, \ldots, s_k\}$ and transition matrix P starts in state s_i, then we can define the hitting time

$$T_{i,j} = \min\{n \geq 1 : X_n = s_j\}$$

with the convention that $T_{i,j} = \infty$ if the Markov chain never visits s_j. We also define the **mean hitting time**

$$\tau_{i,j} = \mathbf{E}[T_{i,j}] \, .$$

This means that $\tau_{i,j}$ is the expected time taken until we come to state s_j, starting from state s_i. For the case $i = j$, we call $\tau_{i,i}$ the **mean return time** for state s_i. We emphasize that when dealing with the hitting time $T_{i,j}$, there is always the implicit assumption that $X_0 = s_i$.

---

[15] By careful modification of our proofs, it is possible to show that Theorem 5.1 holds for arbitrary Markov chains, and that Theorem 5.3 holds without the aperiodicity assumption. That irreducibility and aperiodicity are needed for Theorem 5.2, and irreducibility is needed for Theorem 5.3, will be established by means of counterexamples in Problems 5.2 and 5.3.

Lemma 5.1 *For any irreducible aperiodic Markov chain with state space* $S = \{s_1, \ldots, s_k\}$ *and transition matrix P, we have for any two states* $s_i, s_j \in S$ *that if the chain starts in state* s_i, *then*

$$\mathbf{P}(T_{i,j} < \infty) = 1. \tag{24}$$

Moreover, the mean hitting time $\tau_{i,j}$ *is finite,*[16] *i.e.,*

$$\mathbf{E}[T_{i,j}] < \infty. \tag{25}$$

Proof By Corollary 4.1, we can find an $M < \infty$ such that $(P^M)_{i,j} > 0$ for all $i, j \in \{1, \ldots, k\}$. Fix such an M, set $\alpha = \min\{(P^M)_{i,j} : i, j \in \{1, \ldots, k\}\}$, and note that $\alpha > 0$. Fix two states s_i and s_j as in the lemma, and suppose that the chain starts in s_i. Clearly,

$$\mathbf{P}(T_{i,j} > M) \le \mathbf{P}(X_M \neq s_j) \le 1 - \alpha.$$

Furthermore, given everything that has happened up to time M, we have conditional probability at least α of hitting state s_j at time $2M$, so that

$$\begin{aligned}
\mathbf{P}(T_{i,j} > 2M) &= \mathbf{P}(T_{i,j} > M)\mathbf{P}(T_{i,j} > 2M \mid T_{i,j} > M) \\
&\le \mathbf{P}(T_{i,j} > M)\mathbf{P}(X_{2M} \neq s_j \mid T_{i,j} > M) \\
&\le (1 - \alpha)^2.
\end{aligned}$$

Iterating this argument, we get for any l that

$$\begin{aligned}
\mathbf{P}(T_{i,j} > lM) &= \mathbf{P}(T_{i,j} > M)\mathbf{P}(T_{i,j} > 2M \mid T_{i,j} > M) \cdots \\
&\quad \times \mathbf{P}(T_{i,j} > lM \mid T_{i,j} > (l-1)M) \\
&\le (1 - \alpha)^l,
\end{aligned}$$

which tends to 0 as $l \to \infty$. Hence $\mathbf{P}(T_{i,j} = \infty) = 0$, so (24) is established.

To prove (25), we use the formula (1) for expectation, and get

$$\begin{aligned}
\mathbf{E}[T_{i,j}] &= \sum_{n=1}^{\infty} \mathbf{P}(T_{i,j} \ge n) = \sum_{n=0}^{\infty} \mathbf{P}(T_{i,j} > n) \tag{26} \\
&= \sum_{l=0}^{\infty} \sum_{n=lM}^{(l+1)M-1} \mathbf{P}(T_{i,j} > n)
\end{aligned}$$

[16] If you think that this should follow immediately from (24), then take a look at Example 1.1 to see that things are not always quite that simple.

$$\leq \sum_{l=0}^{\infty} \sum_{n=lM}^{(l+1)M-1} \mathbf{P}(T_{i,j} > lM) = M \sum_{l=0}^{\infty} \mathbf{P}(T_{i,j} > lM)$$

$$\leq M \sum_{l=0}^{\infty} (1-\alpha)^l = M \frac{1}{1-(1-\alpha)} = \frac{M}{\alpha} < \infty.$$

$\square$

Proof of Theorem 5.1 Write, as usual, $(X_0, X_1, \ldots)$ for the Markov chain, $S = \{s_1, \ldots, s_k\}$ for the state space, and P for the transition matrix. Suppose that the chain starts in state s_1, and define, for $i = 1, \ldots, k$,

$$\rho_i = \sum_{n=0}^{\infty} \mathbf{P}(X_n = s_i, T_{1,1} > n)$$

so that in other words, ρ_i is the expected number of visits to state i up to time $T_{1,1} - 1$. Since the mean return time $\mathbf{E}[T_{1,1}] = \tau_{1,1}$ is finite, and $\rho_i < \tau_{1,1}$, we get that ρ_i is finite as well. Our candidate for a stationary distribution is

$$\pi = (\pi_1, \ldots, \pi_k) = \left(\frac{\rho_1}{\tau_{1,1}}, \frac{\rho_2}{\tau_{1,1}}, \ldots, \frac{\rho_k}{\tau_{1,1}} \right).$$

We need to verify that this choice of π satisfies conditions (i) and (ii) of Definition 5.1.

We first show that the relation $\sum_{i=1}^{k} \pi_i P_{i,j} = \pi_j$ in condition (ii) holds for $j \neq 1$ (the case $j = 1$ will be treated separately). We get (hold on!)

$$\pi_j = \frac{\rho_j}{\tau_{1,1}} = \frac{1}{\tau_{1,1}} \sum_{n=0}^{\infty} \mathbf{P}(X_n = s_j, T_{1,1} > n)$$

$$= \frac{1}{\tau_{1,1}} \sum_{n=1}^{\infty} \mathbf{P}(X_n = s_j, T_{1,1} > n) \tag{27}$$

$$= \frac{1}{\tau_{1,1}} \sum_{n=1}^{\infty} \mathbf{P}(X_n = s_j, T_{1,1} > n - 1) \tag{28}$$

$$= \frac{1}{\tau_{1,1}} \sum_{n=1}^{\infty} \sum_{i=1}^{k} \mathbf{P}(X_{n-1} = s_i, X_n = s_j, T_{1,1} > n - 1)$$

$$= \frac{1}{\tau_{1,1}} \sum_{n=1}^{\infty} \sum_{i=1}^{k} \mathbf{P}(X_{n-1} = s_i, T_{1,1} > n - 1) \mathbf{P}(X_n = s_j \mid X_{n-1} = s_i) \tag{29}$$

$$= \frac{1}{\tau_{1,1}} \sum_{n=1}^{\infty} \sum_{i=1}^{k} P_{i,j} \mathbf{P}(X_{n-1} = s_i, T_{1,1} > n - 1)$$

$$= \frac{1}{\tau_{1,1}} \sum_{i=1}^{k} P_{i,j} \sum_{n=1}^{\infty} \mathbf{P}(X_{n-1} = s_i, T_{1,1} > n - 1)$$

$$= \frac{1}{\tau_{1,1}} \sum_{i=1}^{k} P_{i,j} \sum_{m=0}^{\infty} \mathbf{P}(X_m = s_i, T_{1,1} > m)$$

$$= \frac{\sum_{i=1}^{k} \rho_i P_{i,j}}{\tau_{1,1}} = \sum_{i=1}^{k} \pi_i P_{i,j} \tag{30}$$

where in lines (27), (28) and (29) we used the assumption that $j \neq 1$; note also that (29) uses the fact that the event $\{T_{1,1} > n - 1\}$ is determined solely by the variables $X_0, \ldots, X_{n-1}$.

Next, we verify condition (ii) also for the case $j = 1$. Note first that $\rho_1 = 1$; this is immediate from the definition of ρ_i. We get

$$\rho_1 = 1 = \mathbf{P}(T_{1,1} < \infty) = \sum_{n=1}^{\infty} \mathbf{P}(T_{1,1} = n)$$

$$= \sum_{n=1}^{\infty} \sum_{i=1}^{k} \mathbf{P}(X_{n-1} = s_i, T_{1,1} = n)$$

$$= \sum_{n=1}^{\infty} \sum_{i=1}^{k} \mathbf{P}(X_{n-1} = s_i, T_{1,1} > n - 1)\mathbf{P}(X_n = s_1 \mid X_{n-1} = s_i)$$

$$= \sum_{n=1}^{\infty} \sum_{i=1}^{k} P_{i,1}\mathbf{P}(X_{n-1} = s_i, T_{1,1} > n - 1)$$

$$= \sum_{i=1}^{k} P_{i,1} \sum_{n=1}^{\infty} \mathbf{P}(X_{n-1} = s_i, T_{1,1} > n - 1)$$

$$= \sum_{i=1}^{k} P_{i,1} \sum_{m=0}^{\infty} \mathbf{P}(X_m = s_i, T_{1,1} > m)$$

$$= \sum_{i=1}^{k} \rho_i P_{i,1} .$$

Hence

$$\pi_1 = \frac{\rho_1}{\tau_{1,1}} = \sum_{i=1}^{k} \frac{\rho_i P_{i,1}}{\tau_{1,1}} = \sum_{i=1}^{k} \pi_i P_{i,1} .$$

By combining this with (30), we have established that condition (ii) holds for our choice of π.

It remains to show that condition (i) holds as well. That $\pi_i \geq 0$ for $i = 1, \ldots, k$ is obvious. To see that $\sum_{i=1}^k \pi_i = 1$ holds as well, note that

$$\tau_{1,1} = \mathbf{E}[T_{1,1}] = \sum_{n=0}^{\infty} \mathbf{P}(T_{1,1} > n) \tag{31}$$

$$= \sum_{n=0}^{\infty} \sum_{i=1}^{k} \mathbf{P}(X_n = s_i, T_{1,1} > n)$$

$$= \sum_{i=1}^{k} \sum_{n=0}^{\infty} \mathbf{P}(X_n = s_i, T_{1,1} > n)$$

$$= \sum_{i=1}^{k} \rho_i$$

(where equation (31) uses (26)) so that

$$\sum_{i=1}^{k} \pi_i = \frac{1}{\tau_{1,1}} \sum_{i=1}^{k} \rho_i = 1,$$

and condition (i) is verified. $\qquad\square$

We shall go on to consider the asymptotic behavior of the distribution $\mu^{(n)}$ of a Markov chain with arbitrary initial distribution $\mu^{(0)}$. To state the main result (Theorem 5.2), we need to define what it means for a sequence of probability distributions $\nu^{(1)}, \nu^{(2)}, \ldots$ to converge to another probability distribution ν, and to this end it is useful to have a metric on probability distributions. There are various such metrics; one which is useful here is the so-called **total variation distance**.

Definition 5.2 *If $\nu^{(1)} = (\nu_1^{(1)}, \ldots, \nu_k^{(1)})$ and $\nu^{(2)} = (\nu_1^{(2)}, \ldots, \nu_k^{(2)})$ are probability distributions on $S = \{s_1, \ldots, s_k\}$, then we define the **total variation distance** between $\nu^{(1)}$ and $\nu^{(2)}$ as*

$$d_{TV}(\nu^{(1)}, \nu^{(2)}) = \frac{1}{2} \sum_{i=1}^{k} |\nu_i^{(1)} - \nu_i^{(2)}|. \tag{32}$$

If $\nu^{(1)}, \nu^{(2)}, \ldots$ and ν are probability distributions on S, then we say that $\nu^{(n)}$ **converges to ν in total variation as $n \to \infty$**, *writing $\nu^{(n)} \xrightarrow{\text{TV}} \nu$, if*

$$\lim_{n \to \infty} d_{TV}(\nu^{(n)}, \nu) = 0.$$

The constant $\frac{1}{2}$ in (32) is designed to make the total variation distance d_{TV} take values between 0 and 1. If $d_{TV}(\nu^{(1)}, \nu^{(2)}) = 0$, then $\nu^{(1)} = \nu^{(2)}$. In the other extreme case $d_{TV}(\nu^{(1)}, \nu^{(2)}) = 1$, we have that $\nu^{(1)}$ and $\nu^{(2)}$ are "disjoint" in the sense that S can be partitioned into two disjoint subsets S' and S'' such that $\nu^{(1)}$ puts all of its probability mass in S', and $\nu^{(2)}$ puts all of its in S''. The total variation distance also has the natural interpretation

$$d_{TV}(\nu^{(1)}, \nu^{(2)}) = \max_{A \subseteq S} |\nu^{(1)}(A) - \nu^{(2)}(A)|, \tag{33}$$

an identity that you will be asked to prove in Problem 5.1 below. In words, the total variation distance between $\nu^{(1)}$ and $\nu^{(2)}$ is the maximal difference between the probabilities that the two distributions assign to any one event.

We are now ready to state the main result about convergence to stationarity.

Theorem 5.2 (The Markov chain convergence theorem) *Let $(X_0, X_1, \ldots)$ be an irreducible aperiodic Markov chain with state space $S = \{s_1, \ldots, s_k\}$, transition matrix P, and arbitrary initial distribution $\mu^{(0)}$. Then, for any distribution π which is stationary for the transition matrix P, we have*

$$\mu^{(n)} \xrightarrow{\text{TV}} \pi. \tag{34}$$

What the theorem says is that if we run a Markov chain for a sufficiently long time n, then, regardless of what the initial distribution was, the distribution at time n will be close to the stationary distribution π. This is often referred to as the Markov chain approaching **equilibrium** as $n \to \infty$.

For the proof, we will use a so-called **coupling** argument; coupling is one of the most useful and elegant techniques in contemporary probability. Before doing the proof, however, the reader is urged to glance ahead at Theorem 5.3 and its proof, to see how easily Theorem 5.2 implies that there cannot be more than one stationary distribution.

Proof of Theorem 5.2 When studying the behavior of $\mu^{(n)}$, we may assume that $(X_0, X_1, \ldots)$ has been obtained by the simulation method outlined in Chapter 3, i.e.,

$$X_0 = \psi_{\mu^{(0)}}(U_0)$$
$$X_1 = \phi(X_0, U_1)$$
$$X_2 = \phi(X_1, U_2)$$
$$\vdots$$

where $\psi_{\mu^{(0)}}$ is a valid initiation function for $\mu^{(0)}$, ϕ is a valid update function for P, and $(U_0, U_1, \ldots)$ is an i.i.d. sequence of uniform $[0, 1]$ random variables.

Next, we introduce a second Markov chain[17] $(X'_0, X'_1, \ldots)$ by letting ψ_π be a valid initiation function for the distribution π, letting $(U'_0, U'_1, \ldots)$ be another i.i.d. sequence (independent of $(U_0, U_1, \ldots)$) of uniform $[0, 1]$ random variables, and setting

$$X'_0 = \psi_\pi(U_0)$$
$$X'_1 = \phi(X'_0, U'_1)$$
$$X'_2 = \phi(X'_1, U'_2)$$
$$\vdots$$

Since π is a stationary distribution, we have that X'_n has distribution π for any n. Also, the chains $(X_0, X_1, \ldots)$ and $(X'_0, X'_1, \ldots)$ are independent of each other, by the assumption that the sequences $(U_0, U_1, \ldots)$ and $(U'_0, U'_1, \ldots)$ are independent of each other.

A key step in the proof is now to show that, with probability 1, the two chains will "meet", meaning that there exists an n such that $X_n = X'_n$. To show this, define the "first meeting time"

$$T = \min\{n : X_n = X'_n\}$$

with the convention that $T = \infty$ if the chains never meet. Since the Markov chain $(X_0, X_1, \ldots)$ is irreducible and aperiodic, we can find, using Corollary 4.1, an $M < \infty$ such that

$$(P^M)_{i,j} > 0 \text{ for all } i, j \in \{1, \ldots, k\}.$$

Set

$$\alpha = \min\{(P^M)_{i,j} : i \in \{1, \ldots, k\}\},$$

and note that $\alpha > 0$. We get that

$$
\begin{aligned}
\mathbf{P}(T \leq M) &\geq \mathbf{P}(X_M = X'_M) \\
&\geq \mathbf{P}(X_M = s_1, X'_M = s_1) \\
&= \mathbf{P}(X_M = s_1)\mathbf{P}(X'_M = s_1) \\
&= \left(\sum_{i=1}^k \mathbf{P}(X_0 = s_i, X_M = s_1)\right)\left(\sum_{i=1}^k \mathbf{P}(X'_0 = s_i, X'_M = s_1)\right)
\end{aligned}
$$

---

[17] This is what characterizes the coupling method: to construct two or more processes on the same probability space, in order to draw conclusions about their respective distributions.

$$= \left(\sum_{i=1}^{k} \mathbf{P}(X_0 = s_i) \mathbf{P}(X_M = s_1 \mid X_0 = s_i) \right)$$

$$\times \left(\sum_{i=1}^{k} \mathbf{P}(X'_0 = s_i) \mathbf{P}(X'_M = s_1 \mid X'_0 = s_i) \right)$$

$$\geq \left(\alpha \sum_{i=1}^{k} \mathbf{P}(X_0 = s_i) \right) \left(\alpha \sum_{i=1}^{k} \mathbf{P}(X'_0 = s_i) \right) = \alpha^2$$

so that

$$\mathbf{P}(T > M) \leq 1 - \alpha^2 .$$

Similarly, given everything that has happened up to time M, we have conditional probability at least α^2 of having $X_{2M} = X'_{2M} = s_1$, so that

$$\mathbf{P}(X_{2M} \neq X'_{2M} \mid T > M) \leq 1 - \alpha^2 .$$

Hence,

$$\begin{aligned}
\mathbf{P}(T > 2M) &= \mathbf{P}(T > M)\mathbf{P}(T > 2M \mid T > M) \\
&\leq (1 - \alpha^2)\mathbf{P}(T > 2M \mid T > M) \\
&\leq (1 - \alpha^2)\mathbf{P}(X_{2M} \neq X'_{2M} \mid T > M) \\
&\leq (1 - \alpha^2)^2 .
\end{aligned}$$

By iterating this argument, we get for any l that

$$\mathbf{P}(T > lM) \leq (1 - \alpha^2)^l$$

which tends to 0 as $l \to \infty$. Hence,

$$\lim_{n \to \infty} \mathbf{P}(T > n) = 0 \tag{35}$$

so that in other words, we have shown that the two chains will meet with probability 1.

The next step of the proof is to construct a third Markov chain $(X''_0, X''_1, \ldots)$, by setting

$$X''_0 = X_0 \tag{36}$$

and, for each n,

$$X''_{n+1} = \begin{cases} \phi(X''_n, U_{n+1}) & \text{if } X''_n \neq X'_n \\ \phi(X''_n, U'_{n+1}) & \text{if } X''_n = X'_n. \end{cases}$$

In other words, the chain $(X''_0, X''_1, \ldots)$ evolves exactly like the chain $(X_0, X_1, \ldots)$ until the time T when it first meets the chain $(X'_0, X'_1, \ldots)$. It then switches to evolving exactly like the chain $(X_0', X_1', \ldots)$. It is important to realize that $(X_0'', X_1'', \ldots)$ really is a Markov chain with transition matrix P. This may require a pause for thought, but the basic reason why it is true is that at each update, the update function is exposed to a "fresh" new uniform $[0, 1]$ variable, i.e., one which is independent of all previous random variables. (Whether the new chain is exposed to U_{n+1} or to U_{n+1}' depends on the earlier values of the uniform $[0, 1]$ variables, but this does not matter since U_{n+1} and U_{n+1}' have the same distribution and are both independent of everything that has happened up to time n.)

Because of (36), we have that X_0'' has distribution $\mu^{(0)}$. Hence, for any n, X_n'' has distribution $\mu^{(n)}$. Now, for any $i \in \{1, \ldots, k\}$ we get

$$
\begin{aligned}
\mu_i^{(n)} - \pi_i &= \mathbf{P}(X_n'' = s_i) - \mathbf{P}(X_n' = s_i) \\
&\leq \mathbf{P}(X_n'' = s_i, X_n' \neq s_i) \\
&\leq \mathbf{P}(X_n'' \neq X_n') \\
&= \mathbf{P}(T > n)
\end{aligned}
$$

which tends to 0 as $n \to \infty$, due to (35). Using the same argument (with the roles of X_n'' and X_n' interchanged), we see that

$$
\pi_i - \mu_i^{(n)} \leq \mathbf{P}(T > n)
$$

as well, again tending to 0 as $n \to \infty$. Hence,

$$
\lim_{n \to \infty} |\mu_i^{(n)} - \pi_i| = 0 .
$$

This implies that

$$
\begin{aligned}
\lim_{n \to \infty} d_{\mathrm{TV}}(\mu^{(n)}, \pi) &= \lim_{n \to \infty} \left(\tfrac{1}{2} \sum_{i=1}^{k} |\mu_i^{(n)} - \pi_i| \right) \\
&= 0
\end{aligned}
\tag{37}
$$

since each term in the right-hand side of (37) tends to 0. Hence, (34) is established. $\square$

Theorem 5.3 (Uniqueness of the stationary distribution) *Any irreducible and aperiodic Markov chain has exactly one stationary distribution.*

Proof Let $(X_0, X_1, \ldots)$ be an irreducible and aperiodic Markov chain with transition matrix P. By Theorem 5.1, there exists *at least* one stationary distribution for P, so we only need to show that there is *at most* one stationary distribution. Let π and π' be two (a priori possibly different) stationary distributions for P; our task is to show that $\pi = \pi'$.

Suppose that the Markov chain starts with initial distribution $\mu^{(0)} = \pi'$. Then $\mu^{(n)} = \pi'$ for all n, by the assumption that π' is stationary. On the other hand, Theorem 5.2 tells us that $\mu^{(n)} \xrightarrow{\text{TV}} \pi$, meaning that

$$\lim_{n\to\infty} d_{\text{TV}}(\mu^{(n)}, \pi) = 0.$$

Since $\mu^{(n)} = \pi'$, this is the same as

$$\lim_{n\to\infty} d_{\text{TV}}(\pi', \pi) = 0.$$

But $d_{\text{TV}}(\pi', \pi)$ does not depend on n, and hence equals 0. This implies that $\pi = \pi'$, so the proof is complete. $\square$

To summarize Theorems 5.2 and 5.3: If a Markov chain is irreducible and aperiodic, then it has a unique stationary distribution π, and the distribution $\mu^{(n)}$ of the chain at time n approaches π as $n \to \infty$, regardless of the initial distribution $\mu^{(0)}$.

Problems

5.1 **(7)** Prove the formula (33) for total variation distance. Hint: consider the event

$$A = \{s \in S : v^{(1)}(s) \geq v^{(2)}(s)\}.$$

5.2 **(4) Theorems 5.2 and 5.3 fail for reducible Markov chains.** Consider the reducible Markov chain in Example 4.1.
   (a) Show that both $\pi = (0.375, 0.625, 0, 0)$ and $\pi' = (0, 0, 0.5, 0.5)$ are stationary distributions for this Markov chain.
   (b) Use (a) to show that the conclusions of Theorem 5.2 and 5.3 fail for this Markov chain.

5.3 **(6) Theorem 5.2 fails for periodic Markov chains.** Consider the Markov chain $(X_0, X_1, \ldots)$ describing a knight making random moves on a chessboard, as in Problem 4.3 (c). Show that $\mu^{(n)}$ does not converge in total variation, if the chain is started in a fixed state (such as the square a1 of the chessboard).

5.4 **(7) If there are two different stationary distributions, then there are infinitely many.** Suppose that $(X_0, X_1, \ldots)$ is a reducible Markov chain with two different stationary distributions π and π'. Show that, for any $p \in (0, 1)$, we get yet another stationary distribution as $p\pi + (1 - p)\pi'$.

5.5 **(6)** Show that the stationary distribution obtained in the proof of Theorem 5.1 can be written as

$$\pi = \left(\frac{1}{\tau_{1,1}}, \frac{1}{\tau_{2,2}}, \ldots, \frac{1}{\tau_{k,k}}\right).$$

6

Reversible Markov chains

In this chapter we introduce a special class of Markov chains known as the **reversible** ones. They are called so because they, in a certain sense, look the same regardless of whether time runs backwards or forwards; this is made precise in Problem 6.3 below. Such chains arise naturally in the algorithmic applications of Chapters 7–13, as well as in several other applied contexts. We jump right on to the definition:

Definition 6.1 *Let* $(X_0, X_1, \ldots)$ *be a Markov chain with state space* $S = \{s_1, \ldots, s_k\}$ *and transition matrix* P. *A probability distribution* π *on* S *is said to be* **reversible** *for the chain (or for the transition matrix* P) *if for all* $i, j \in \{1, \ldots, k\}$ *we have*

$$\pi_i P_{i,j} = \pi_j P_{j,i} \,. \tag{38}$$

The Markov chain is said to be reversible if there exists a reversible distribution for it.

If the chain is started with the reversible distribution π, then the left-hand side of (38) can be thought of as the amount of probability mass flowing at time 1 from state s_i to state s_j. Similarly, the right-hand side is the probability mass flowing from s_j to s_i. This seems like (and is!) a strong form of equilibrium, and the following result suggests itself.

Theorem 6.1 *Let* $(X_0, X_1, \ldots)$ *be a Markov chain with state space* $S = \{s_1, \ldots, s_k\}$ *and transition matrix* P. *If* π *is a reversible distribution for the chain, then it is also a stationary distribution for the chain.*

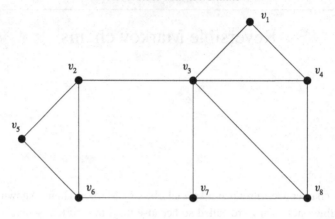

Fig. 4. A graph.

Proof Property (i) of Definition 5.1 is immediate, so it only remains to show that for any $j \in \{1, \ldots, k\}$, we have

$$\pi_j = \sum_{i=1}^{k} \pi_i P_{i,j}.$$

We get

$$\pi_j = \pi_j \sum_{i=1}^{k} P_{j,i} = \sum_{i=1}^{k} \pi_j P_{j,i} = \sum_{i=1}^{k} \pi_i P_{i,j},$$

where the last equality uses (38). $\square$

We go on to consider some examples.

> **Example 6.1: Random walks on graphs.** This example is a generalization of the random walk example in Figure 1. A **graph** $G = (V, E)$ consists of a **vertex set** $V = \{v_1, \ldots, v_k\}$, together with an **edge set** $E = \{e_1, \ldots, e_l\}$. Each edge connects two of the vertices; an edge connecting the vertices v_i and v_j is denoted $\langle v_i, v_j \rangle$. No two edges are allowed to connect the same pair of vertices. Two vertices are said to be **neighbors** if they share an edge.
>
> For instance, the graph in Figure 4 has vertex set $V = \{v_1, \ldots, v_8\}$ and edge set
>
> $$\begin{aligned} E = \ &\{\langle v_1, v_3 \rangle, \langle v_1, v_4 \rangle, \langle v_2, v_3 \rangle, \langle v_2, v_5 \rangle, \langle v_2, v_6 \rangle, \langle v_3, v_4 \rangle, \\ &\langle v_3, v_7 \rangle, \langle v_3, v_8 \rangle, \langle v_4, v_8 \rangle, \langle v_5, v_6 \rangle, \langle v_6, v_7 \rangle, \langle v_7, v_8 \rangle\}. \end{aligned}$$
>
> A random walk on a graph $G = (V, E)$ is a Markov chain with state space $V = \{v_1, \ldots, v_k\}$ and the following transition mechanism: If the random walker stands at a vertex v_i at time n, then it moves at time $n + 1$ to one of the neighbors of v_i chosen at random, with equal probability for each of the neighbors. Thus, if

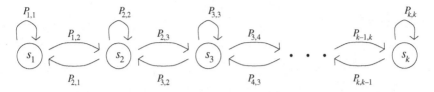

Fig. 5. Transition graph of a Markov chain of the kind discussed in Example 6.2.

we denote the number of neighbors of a vertex v_i by d_i, then the elements of the transition matrix are given by

$$P_{i,j} = \begin{cases} \frac{1}{d_i} & \text{if } v_i \text{ and } v_j \text{ are neighbors} \\ 0 & \text{otherwise.} \end{cases}$$

It turns out that random walks on graphs are reversible Markov chains, with reversible distribution π given by

$$\pi = \left(\frac{d_1}{d}, \frac{d_2}{d}, \dots, \frac{d_k}{d} \right) \tag{39}$$

where $d = \sum_{i=1}^{k} d_i$. To see that (38) holds for this choice of π, we calculate

$$\pi_i P_{i,j} = \begin{cases} \frac{d_i}{d}\frac{1}{d_i} = \frac{1}{d} = \frac{d_j}{d}\frac{1}{d_j} = \pi_j P_{j,i} & \text{if } v_i \text{ and } v_j \text{ are neighbors} \\ 0 = \pi_j P_{j,i} & \text{otherwise.} \end{cases}$$

For the graph in Figure 4, (39) becomes

$$\pi = \left(\frac{2}{24}, \frac{3}{24}, \frac{5}{24}, \frac{3}{24}, \frac{2}{24}, \frac{3}{24}, \frac{3}{24}, \frac{3}{24} \right)$$

so that in equilibrium, v_3 is the most likely vertex for the random walker to be at, whereas v_1 and v_5 are the least likely ones.

Example 6.2 Let $(X_0, X_1, \dots)$ be a Markov chain with state space $S = \{s_1, \dots, s_k\}$ and transition matrix P, and suppose that the transition matrix has the properties that

(i) $P_{i,j} > 0$ whenever $|i - j| = 1$, and
(ii) $P_{i,j} = 0$ whenever $|i - j| \geq 2$.

Such a Markov chain is often called a **birth-and-death process**, and its transition graph has the form outlined in Figure 5 (with some or all of the $P_{i,i}$-"loops" possibly being absent). We claim that any Markov chain of this kind is reversible. To construct a reversible distribution π for the chain, we begin by setting π_1^* equal to some arbitrary strictly positive number a. The condition (38) with $i = 1$ and $j = 2$ forces us to take

$$\pi_2^* = \frac{a P_{1,2}}{P_{2,1}}.$$

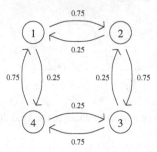

Fig. 6. Transition graph of the Markov chain in Example 6.3.

Applying (38) again, now with $i = 2$ and $j = 3$, we get

$$\pi_3^* = \frac{\pi_2^* P_{2,3}}{P_{3,2}} = \frac{a P_{1,2} P_{2,3}}{P_{2,1} P_{3,2}}.$$

We can continue in this way, and get

$$\pi_i^* = \frac{a \prod_{l=1}^{i-1} P_{l,l+1}}{\prod_{l=1}^{i-1} P_{l+1,l}}$$

for each i. Then $\pi^* = (\pi_1^*, \ldots, \pi_k^*)$ satisfies the requirements of a reversible distribution, except possibly that the entries do not sum to 1, as is required for any probability distribution. But this is easily taken care of by dividing all entries by their sum. It is readily checked that

$$\pi = (\pi_1, \pi_2, \ldots, \pi_k) = \left(\frac{\pi_1^*}{\sum_{i=1}^k \pi_i^*}, \frac{\pi_2^*}{\sum_{i=1}^k \pi_i^*}, \ldots, \frac{\pi_k^*}{\sum_{i=1}^k \pi_i^*} \right)$$

is a reversible distribution.

Having come this far, one might perhaps get the impression that most Markov chains are reversible. This is not really true, however, and to make up for this false impression, let us also consider an example of a Markov chain which is *not* reversible.

Example 6.3: A nonreversible Markov chain. Let us consider a modified version of the random walk in Figure 1. Suppose that the coin tosses used by the random walker in Figure 1 are *biased*, in such a way that at each integer time, he moves one step clockwise with probability $\frac{3}{4}$, and one step counterclockwise with probability $\frac{1}{4}$. This yields a Markov chain with the transition graph in Figure 6. It is clear that $\pi = (\frac{1}{4}, \frac{1}{4}, \frac{1}{4}, \frac{1}{4})$ is a stationary distribution for this chain (right?). Furthermore, since the chain is irreducible, we have by Theorem 5.3 and Footnote 15 in Chapter 5 that this is the only stationary distribution. Because of

Theorem 6.1 we therefore need π to be reversible in order for the Markov chain to be reversible. But if we, e.g., try (38) with $i = 1$ and $j = 2$, we get

$$\pi_1 P_{1,2} = \frac{1}{4} \cdot \frac{3}{4} = \frac{3}{16} > \frac{1}{16} = \frac{1}{4} \cdot \frac{1}{4} = \pi_2 P_{2,1}$$

so that $\pi_1 P_{1,2} \neq \pi_2 P_{2,1}$, and reversibility fails. Intuitively, the reason why this chain is not reversible is that the walker has a tendency to move clockwise. If we filmed the walker and watched the movie backwards, it would look as if he preferred to move counterclockwise, so that in other words the chain behaves differently in "backwards time" compared to "forwards time".

Let us close this chapter by mentioning the existence of a simple and beautiful equivalence between reversible Markov chains on the one hand, and resistor networks on the other. This makes electrical arguments (such as the series and parallel laws) useful for analyzing Markov chains, and conversely, probabilistic argument available in the study of electrical networks. Unfortunately, a discussion of this topic would take us too far, considering the modest format of these notes. Suggestions for further reading can be found in Chapter 14.

Problems

6.1 **(6) The king on the chessboard.** Recall from Problem 4.3 (a) the king making random moves on a chessboard. If you solved that problem correctly, then you know that the corresponding Markov chain is irreducible and aperiodic. By Theorem 5.3, the chain therefore has a unique stationary distribution π. Compute π. Hint: the chain is reversible, and can be handled as in Example 6.1.

6.2 **(8) Ehrenfest's urn model.** Fix an integer k, and imagine two urns, each containing a number of balls, in such a way that the total number of balls in the two urns is k. At each integer time, we pick one ball at random (each with probability $\frac{1}{k}$) and move it to the other urn.[18] If X_n denotes the number of balls in the first urn, then $(X_0, X_1, \ldots)$ forms a Markov chain with state space $\{0, \ldots, k\}$.

(a) Write down the transition matrix of this Markov chain.

(b) Show that the Markov chain is reversible with stationary distribution π given by

$$\pi_i = \frac{k!}{i!(k-i)!} 2^{-k} \quad \text{for } i = 0, \ldots, k.$$

(c) Show that the same distribution (known as the **binomial distribution**) also arises as the distribution of a binomial $(k, \frac{1}{2})$ random variable, as defined in Example 1.3.

(d) Can you give an intuitive explanation of why Ehrenfest's urn model and Example 1.3 give rise to the same distribution?

[18] There are various interpretations of this model. Ehrenfest's original intention was to model diffusion of molecules between the two halves of a gas container.

6.3 (7) **Time reversal.** Let $(X_0, X_1, \ldots)$ be a reversible Markov chain with state space S, transition matrix P, and reversible distribution π. Show that if the chain is started with initial distribution $\mu^{(0)} = \pi$, then for any n and any $s_{i_0}, s_{i_1}, \ldots, s_{i_n} \in S$, we have

$$\mathbf{P}(X_0 = s_{i_0}, X_1 = s_{i_1}, \ldots, X_n = s_{i_n}) = \mathbf{P}(X_0 = s_{i_n}, X_1 = s_{i_{n-1}}, \ldots, X_n = s_{i_0}).$$

In other words, the chain is equally likely to make a tour through the states $s_{i_0}, \ldots s_{i_n}$ in forwards as in backwards order.

7

Markov chain Monte Carlo

In this chapter and the next, we consider the following problem: Given a probability distribution π on $S = \{s_1, \ldots, s_k\}$, how do we simulate a random object with distribution π? To motivate the problem, we begin with an example.

Example 7.1: The hard-core model. Let $G = (V, E)$ be a graph (recall Example 6.1 for the definition of a graph) with vertex set $V = \{v_1, \ldots, v_k\}$ and edge set $E = \{e_1, \ldots, e_l\}$. In the so-called hard-core model on G, we randomly assign the value 0 or 1 to each of the vertices, in such a way that no two adjacent vertices (i.e., no two vertices that share an edge) both take the value 1. Assignments of 0's and 1's to the vertices are called **configurations**, and can be thought of as elements of the set $\{0, 1\}^V$. Configurations in which no two 1's occupy adjacent vertices are called **feasible**. The precise way in which we pick a random configuration is to take each of the feasible configurations with equal probability. We write μ_G for the resulting probability measure on $\{0, 1\}^V$. Hence, for $\xi \in \{0, 1\}^V$, we have

$$\mu_G(\xi) = \begin{cases} \frac{1}{Z_G} & \text{if } \xi \text{ is feasible} \\ 0 & \text{otherwise,} \end{cases} \tag{40}$$

where Z_G is the total number of feasible configurations for G. See Figure 7 for a random configuration chosen according to μ_G in the case where G is a square grid of size 8×8.

This model (with the graph G being a three-dimensional grid) was introduced in statistical physics to capture some of the behavior of a gas whose particles have nonnegligible radii and cannot overlap; here 1's represent particles and 0's represent empty locations. (The model has also been used in telecommunications for modelling situations where an occupied node disables all its neighboring nodes.)

A very natural question is now: What is the expected number of 1's in a random configuration chosen according to μ_G? If we write $n(\xi)$ for the number of 1's in the configuration ξ, and X for a random configuration chosen according to μ_G,

45

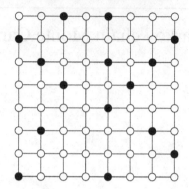

Fig. 7. A feasible configuration (chosen at random according to the probability measure μ_G), where G is a square grid of size 8×8. Black (resp. white) circles represent 1's (resp. 0's). Note that no two 1's occupy adjacent vertices.

then this expected value is given by

$$\mathbf{E}[n(X)] = \sum_{\xi \in \{0,1\}^V} n(\xi)\mu_G(\xi) = \frac{1}{Z_G} \sum_{\xi \in \{0,1\}^V} n(\xi)\mathbf{I}_{\{\xi \text{ is feasible}\}}, \qquad (41)$$

where Z_G is the total number of feasible configurations for the graph G. To evaluate this sum may be infeasible unless the graph is very small, since the number of configurations (and hence the number of terms in the sum) grows exponentially in the size of the graph (for instance, we get $2^{64} \approx 1.8 \cdot 10^{19}$ different configurations for the moderately-sized graph in Figure 7; in physical applications one is usually interested in much bigger graphs). It may help somewhat that most of the terms take the value 0, but the number of nonzero terms grows exponentially as well. Note also that the calculation of Z_G is computationally nontrivial.

When the exact expression in (41) is beyond what our computational resources can deal with, a good idea may be to revert to simulations. If we know how to simulate a random configuration X with distribution μ_G, then we can do this many times, and estimate $\mathbf{E}[n(X)]$ by the average number of 1's in our simulations. By the Law of Large Numbers (Theorem 1.2), this estimate converges to the true value of $\mathbf{E}[n(X)]$, as the number of simulations tends to infinity, and we can form confidence intervals etc., using standard statistical procedures.

With this example in mind, let us discuss how we can simulate a random variable X distributed according to a given probability distribution π on a state space S. In principle it is very simple: just enumerate the elements of S as $s_1, \ldots, s_k$, and then let

$$X = \psi(U)$$

where U is a uniform $[0, 1]$ random variable, and the function $\psi : [0, 1] \to S$ is given by

$$
\psi(x) = \begin{cases}
s_1 & \text{for } x \in [0, \pi(s_1)) \\
s_2 & \text{for } x \in [\pi(s_1), \pi(s_1) + \pi(s_2)) \\
\vdots & \quad \vdots \\
s_i & \text{for } x \in \left[\sum_{j=1}^{i-1} \pi(s_j), \sum_{j=1}^{i} \pi(s_j) \right) \\
\vdots & \quad \vdots \\
s_k & \text{for } x \in \left[\sum_{j=1}^{k-1} \pi(s_j), 1 \right]
\end{cases} \tag{42}
$$

as in (18). By arguing as in Chapter 3, we see that this gives X the desired distribution π. In practice, however, this approach is infeasible unless the state space S is small. For the hard-core model on a square grid the size of a chessboard or bigger, the evaluation of the function ψ in (42) becomes too time-consuming for this method to be of any practical use.

It is precisely in this kind of situation that the **Markov chain Monte Carlo (MCMC)** method is useful. The method originates in physics, where the earliest uses go back to the 1950's. It later enjoyed huge booms in other areas, especially in image analysis in the 1980's, and in the increasingly important area of statistics known as **Bayesian statistics**[19] in the 1990's.

The idea is the following: Suppose we can construct an irreducible and aperiodic Markov chain $(X_0, X_1, \ldots)$, whose (unique) stationary distribution is π. If we run the chain with arbitrary initial distribution (for instance, starting in a fixed state), then the Markov chain convergence theorem (Theorem 5.2) guarantees that the distribution of the chain at time n converges to π, as $n \to \infty$. Hence, if we run the chain for a sufficiently long time n, then the distribution of X_n will be very close to π. Of course this is just an approximation, but the point is that the approximation can be made arbitrarily good by picking the running time n large.

A natural objection at this stage is: How can it possibly be any easier to construct a Markov chain with the desired property than to construct a random variable with distribution π directly? To answer this, we move straight on to an example.

Example 7.2: An MCMC algorithm for the hard-core model. Let us consider the hard-core model in Example 7.1 on a graph $G = (V, E)$ (which for concreteness may be taken to be the one in Figure 7) with $V = \{v_1, \ldots, v_k\}$. In order

---

[19] In fact, it may be argued that the main reason that the Bayesian approach to statistics has gained ground compared to classical (frequentist) statistics is that MCMC methods have provided the computational tool that makes the approach feasible in practice.

to get an MCMC algorithm for this model, we want to construct a Markov chain whose state space S is the set of feasible configurations for G, i.e.,

$$S = \{\xi \in \{0, 1\}^V : \xi \text{ is feasible}\}. \tag{43}$$

In addition, we want the Markov chain to be irreducible and aperiodic, and have stationary distribution μ_G given by (40).

A Markov chain $(X_0, X_1, \ldots)$ with the desired properties can be obtained using the following transition mechanism. At each integer time $n + 1$, we do as follows:

1. Pick a vertex $v \in V$ at random (uniformly).
2. Toss a fair coin.
3. If the coin comes up heads, and all neighbors of v take value 0 in X_n, then let $X_{n+1}(v) = 1$; otherwise let $X_{n+1}(v) = 0$.
4. For all vertices w other than v, leave the value at w unchanged, i.e., let $X_{n+1}(w) = X_n(w)$.

It is not difficult to verify that this Markov chain is irreducible and aperiodic; see Problem 7.1. Hence, it just remains to show that μ_G is a stationary distribution for the chain. By Theorem 6.1, it is enough to show that μ_G is reversible. Letting $P_{\xi,\xi'}$ denote the transition probability from state ξ to state ξ' (with transition mechanism as above), we thus need to check that

$$\mu_G(\xi) P_{\xi,\xi'} = \mu_G(\xi') P_{\xi',\xi} \tag{44}$$

for any two feasible configurations ξ and ξ'. Let us write $d = d(\xi, \xi')$ for the number of vertices in which ξ and ξ' differ, and treat the three cases $d = 0, d = 1$ and $d \geq 2$ separately. Firstly, the case $d = 0$ means that $\xi = \xi'$, in which case the relation (44) is completely trivial. Secondly, the case $d \geq 2$ is almost as trivial, because the chain never changes the values at more than one vertex at a time, making both sides of (44) equal to 0. Finally, consider the case $d = 1$ where ξ and ξ' differ at exactly one vertex v. Then all neighbors of v must take the value 0 in both ξ and ξ', since otherwise one of the configurations would not be feasible. We therefore get

$$\mu_G(\xi) P_{\xi,\xi'} = \frac{1}{Z_G} \frac{1}{2k} = \mu_G(\xi') P_{\xi',\xi}$$

and (44) is verified (recall that k is the number of vertices). Hence the chain has μ_G as a reversible (and therefore stationary) distribution.

We can now simulate this Markov chain using the methods of Chapter 3. A convenient choice of update function ϕ is to split the unit interval $[0, 1]$ into $2k$ subintervals of equal length $\frac{1}{2k}$, representing the choices

$$(v_1, \text{ heads}), (v_1, \text{ tails}), (v_2, \text{ heads}), \ldots, (v_k, \text{ tails})$$

in the above description of the transition mechanism. If we now run the chain for a long time n, starting with an arbitrary feasible initial configuration such as the "all 0's" configuration, and output X_n, then we get a random configuration whose distribution is approximately μ_G.

The above is a typical MCMC algorithm in several respects. Firstly, note that although it is only required that the chain has the desired distribution as a stationary distribution, we found a chain with the stronger property that the distribution is reversible. This is the case for the vast majority of known MCMC algorithms. The reason for this is that in most nontrivial situations, the easiest way to construct a chain with a given stationary distribution π is to make sure that the reversibility condition (38) holds.

Secondly, the algorithm in Example 7.2 is an example of a commonly used special class of MCMC algorithms known as **Gibbs samplers**. These are useful to simulate probability distributions π on state spaces of the form S^V, where S and V are finite sets. In other words, we have a finite set V of vertices with a finite set S of attainable values at each vertex, and π is the distribution of some random assignment of values in S to the vertices in V (in the hard-core model example, we have $S = \{0, 1\}$). The Gibbs sampler is a Markov chain which at each integer time $n + 1$ does the following.

1. Pick a vertex $v \in V$ at random (uniformly).
2. Pick $X_{n+1}(v)$ according to the conditional π-distribution of the value at v given that all other vertices take values according to X_n.
3. Let $X_{n+1}(w) = X_n(w)$ for all vertices $w \in V$ except v.

It is not hard to show that this Markov chain is aperiodic, and that it has π as a reversible (hence stationary) distribution. If in addition the chain is irreducible (which may or may not be the case depending on which elements of S^V have nonzero π-probability), then this Markov chain is a correct MCMC algorithm for simulating random variables with distribution π. We give another example:

> **Example 7.3: An MCMC algorithm for random q-colorings.** Let $G = (V, E)$ be a graph, and let $q \geq 2$ be an integer. A q-coloring of the graph G is an assignment of values from $\{1, \ldots, q\}$ (thought of as q different "colors") with the property that no two adjacent vertices have the same value (color). By a random q-coloring for G, we mean a q-coloring chosen uniformly from the set of possible q-colorings for G, and we write $\rho_{G,q}$ for the corresponding probability distribution[20] on S^V.
>
> For a vertex $v \in V$ and an assignment ξ of colors to the vertices other than v, the conditional $\rho_{G,q}$-distribution of the color at v is uniform over the set of all colors that are not attained in ξ at some neighbor of v (right?). A Gibbs sampler

---

[20] We are here making the implicit assumption that there exists at least one q-coloring for G. This is not always the case. For instance, if $q = 2$ and G consists of three vertices connected in a triangle, then no q-coloring can be found. In general it is a difficult combinatorial problem to determine whether q-colorings exist for a given choice of G and q. The famous **four-color theorem** states that if G is a planar graph (i.e., G is a graph that can be drawn in the plane in such a way that no two edges cross each other), then $q = 4$ is enough.

for random q-colorings is therefore an S^V-valued Markov chain where at each time $n + 1$, transitions take place as follows.

1. Pick a vertex $v \in V$ at random (uniformly).
2. Pick $X_{n+1}(v)$ according to the uniform distribution over the set of colors that are not attained at any neighbor of v.
3. Leave the color unchanged at all other vertices, i.e., let $X_{n+1}(w) = X_n(w)$ for all vertices $w \in V$ except v.

This chain is aperiodic and has $\rho_{G,q}$ as a stationary distribution; see Problem 7.3. Whether or not the chain is irreducible depends on G and q, and it is a nontrivial problem in general to determine this.[21] In case we can show that it is irreducible, this Gibbs sampler becomes a useful MCMC algorithm.

Let us also mention that a commonly used variant of the Gibbs sampler is the following. Instead of picking the vertices to update at random, we can cycle systematically through the vertex set. For instance, if $V = \{v_1, \ldots, v_k\}$, we may decide to update vertex

$$\begin{cases} v_1 & \text{at times } 1, k + 1, 2k + 1, \ldots \\ v_2 & \text{at times } 2, k + 2, 2k + 2, \ldots \\ \vdots & \quad \vdots \\ v_i & \text{at times } i, k + i, 2k + i, \ldots \\ \vdots & \quad \vdots \\ v_k & \text{at times } k, 2k, 3k, \ldots . \end{cases} \tag{45}$$

This gives an inhomogeneous Markov chain (because there are k different update rules used at different times) which is aperiodic and has the desired distribution as a reversible distribution. Furthermore, it is irreducible if and only if the original "random vertex" Gibbs sampler is irreducible. To prove these claims is reasonably straightforward, but requires a notationally somewhat inconvenient extension of the theory in Chapters 4–6 to the case of inhomogeneous Markov chains; we therefore omit the details. This variant of the Gibbs sampler is referred to as the **systematic sweep Gibbs sampler**.

Another important general procedure for designing a reversible Markov chain for MCMC algorithms is the construction of a so-called **Metropolis chain**.[22] Let us describe a way (not the most general possible) to construct a Metropolis chain for simulating a given probability distribution $\pi = (\pi_1, \ldots, \pi_k)$ on a set $S = \{s_1, \ldots, s_k\}$. The first step is to construct some

---

[21] Compare with the previous footnote. One thing which is not terribly hard is to show that for any given graph G, the chain is irreducible for all sufficiently large q.

[22] A more general (and widely used) class of Markov chains for MCMC simulation is that of the so-called **Metropolis–Hastings chains**; see the book [GRS] mentioned in Chapter 14.

graph G with vertex set S. The edge set (neighborhood structure) of this graph may be arbitrary, except that

(i) the graph must be connected in order to assure irreducibility of the resulting chain, and

(ii) each vertex should not be the endpoint of too many edges, since otherwise the chain becomes computationally too heavy to simulate in practice.

As usual, we say that two states s_i and s_j are neighbors if the graph contains an edge $\langle s_i, s_j \rangle$ linking them. We also write d_i for the number of neighbors of state s_i. The Metropolis chain corresponding to a given choice of G has transition matrix

$$
P_{i,j} = \begin{cases} \frac{1}{d_i} \min\{\frac{\pi_j d_i}{\pi_i d_j}, 1\} & \text{if } s_i \text{ and } s_j \text{ are neighbors} \\ 0 & \text{if } s_i \neq s_j \text{ are not neighbors} \\ 1 - \sum_{\substack{l \\ s_l \sim s_i}} \frac{1}{d_i} \min\{\frac{\pi_l d_i}{\pi_i d_l}, 1\} & \text{if } i = j , \end{cases} \tag{46}
$$

where the sum is over all states s_l that are neighbors of s_i. This transition matrix corresponds to the following transition mechanism: Suppose that $X_n = s_i$. First pick a state s_j according to uniform distribution on the set of neighbors of s_i (so that each neighbor is chosen with probability $\frac{1}{d_i}$). Then set

$$
X_{n+1} = \begin{cases} s_j & \text{with probability } \min\{\frac{\pi_j d_i}{\pi_i d_j}, 1\} \\ s_i & \text{with probability } 1 - \min\{\frac{\pi_j d_i}{\pi_i d_j}, 1\} . \end{cases}
$$

To show that this Markov chain has π as its stationary distribution, it is enough to verify that the reversibility condition

$$
\pi_i P_{i,j} = \pi_j P_{j,i} \tag{47}
$$

holds for all i and j. We proceed as in Example 7.2, by first noting that (47) is trivial for $i = j$. For the case where $i \neq j$ and s_i and s_j are not neighbors, (47) holds because both sides are 0. Finally, we split the case where s_i and s_j are neighbors into two subcases according to whether or not $\frac{\pi_j d_i}{\pi_i d_j} \geq 1$. If $\frac{\pi_j d_i}{\pi_i d_j} \geq 1$, then

$$
\begin{cases} \pi_i P_{i,j} = \pi_i \frac{1}{d_i} \\ \pi_j P_{j,i} = \pi_j \frac{1}{d_j} \frac{\pi_i d_j}{\pi_j d_i} = \frac{\pi_i}{d_i} \end{cases}
$$

so that (47) holds. Similarly, if $\frac{\pi_j d_i}{\pi_i d_j} < 1$, then

$$\begin{cases} \pi_i P_{i,j} = \pi_i \frac{1}{d_i} \frac{\pi_j d_i}{\pi_i d_j} = \frac{\pi_j}{d_j} \\ \pi_j P_{j,i} = \pi_j \frac{1}{d_j} \end{cases}$$

and again (47) holds. So π is a reversible (hence stationary) distribution for the Metropolis chain, which therefore can be used for MCMC simulation of π.

Problems

7.1 **(5)** Show that the Markov chain used for MCMC simulation of the hard-core model in Example 7.2 is

(a) irreducible,[23] and

(b) aperiodic.[24]

7.2 **(8*)** Write a computer program, using the algorithm in Example 7.2, for simulating the hard-core model on a general $k \times k$ square grid. Then do some simulation experiments.[25]

7.3 **(7)** Show, by arguing as in Example 7.2 and Problem 7.1 (b), that the Gibbs sampler for random q-colorings in Example 7.3

(a) has $\rho_{G,q}$ as a stationary distribution, and

(b) is aperiodic.

7.4 **(6) A generalized hard-core model.** A natural generalization of the hard-core model is to allow for different "packing intensities" of 1's in the graph. This is done by introducing a parameter $\lambda > 0$, and changing the probability measure μ_G defined in (40) into a probability measure $\mu_{G,\lambda}$ in which each configuration $\xi \in \{0, 1\}^V$ gets probability

$$\mu_{G,\lambda}(\xi) = \begin{cases} \frac{\lambda^{n(\xi)}}{Z_{G,\lambda}} & \text{if } \xi \text{ is feasible} \\ 0 & \text{otherwise,} \end{cases} \qquad (48)$$

where $n(\xi)$ is the number of 1's in ξ, and $Z_{G,\lambda} = \sum_{\xi \in \{0,1\}^V} \lambda^{n(\xi)} \mathbf{I}_{\{\xi \text{ is feasible}\}}$ is a normalizing constant. As follows from a direct calculation, this means that for

---

[23] Hint: We need to show that for any two feasible configurations ξ and ξ', the chain can go from ξ to ξ' in a finite number of steps. The easiest way to show this is to demonstrate that it can go from ξ to the "all 0's" configuration in a finite number of steps, and then from the "all 0's" configuration to ξ' in a finite number of steps.

[24] Hint: To show that the period of a state ξ is 1, it is enough to show that the Markov chain can go from ξ to ξ in one step (see also Problem 4.2).

[25] When you have managed to do this for, say, a 10×10 square lattice, consider the following: Think back to Example 2.3 (the Internet as a Markov chain). Did that seem to have a ridiculously huge state space? Well, you have just simulated a Markov chain whose state space is even bigger! It is not hard to show that the state space S as defined in (43) contains at least $2^{k^2/2} = 2^{50} \approx 1.1 \cdot 10^{15}$ elements – much larger than the number of web pages on the Internet today.

any vertex $v \in V$, the conditional probability that v takes the value 1, given the values at all other vertices, equals

$$\begin{cases} \frac{\lambda}{\lambda+1} & \text{if all neighbors of } v \text{ take value } 0 \\ 0 & \text{otherwise.} \end{cases}$$

The model's "desire" to put a 1 at v therefore increases gradually as λ increases from 0 to ∞. The case $\lambda = 1$ reduces to the standard hard-core model in Example 7.1.

Construct an MCMC algorithm for this generalized hard-core model.

8

Fast convergence of MCMC algorithms

Although the MCMC approach to simulation, described in the previous chapter, is highly useful, let us note two drawbacks of the method:

(A) The main theoretical basis for the MCMC method is Theorem 5.2, which guarantees that the distribution $\mu^{(n)}$ at time n of an irreducible and aperiodic Markov chain started in a fixed state converges to the stationary distribution π as $n \to \infty$. But this does not imply that $\mu^{(n)}$ ever becomes *equal* to π, only that it comes very close. As a matter of fact, in the vast majority of examples, we have $\mu^{(n)} \neq \pi$ for all n (see, e.g., Problem 2.3). Hence, no matter how large n is taken to be in the MCMC algorithm, there will still be some discrepancy between the distribution of the output and the desired distribution π.

(B) In order to make the error due to (A) small, we need to figure out how large n needs to be taken in order to make sure that the discrepancy between $\mu^{(n)}$ and π (measured in the total variation distance $d_{TV}(\mu^{(n)}, \pi)$) is smaller than some given $\varepsilon > 0$. In many situations, it has turned out to be very difficult to obtain upper bounds on how large n needs to be taken, that are small enough to be of any practical use.[26]

Problem (A) above is in itself not a particularly serious obstacle. In most situations, we can tolerate a small error in the distribution of the output, as long as we have an idea about how small it is. It is only in combination with (B) that it becomes really bothersome. Due to difficulties in determining the rate of

---

[26] In general, it is possible to extract an explicit upper bound (depending on ε and on the chain) by a careful analysis of the proof of Theorem 5.2. However, this often leads to bounds of astronomical magnitude, such as "$d_{TV}(\mu^{(n)}, \pi) < 0.01$ whenever $n \geq 10^{100}$". This is of course totally useless, because the simulation of 10^{100} steps of a Markov chain is unlikely to terminate during our lifetimes. In such situations, one can often suspect that the convergence is much faster (so that perhaps $n = 10^5$ would suffice), but to actually prove this often turns out to be prohibitively difficult.

convergence to stationarity in Markov chains, much of today's MCMC practice has the following character: A Markov chain $(X_0, X_1, \ldots)$ whose distribution $\mu^{(n)}$ converges to the desired distribution π, as the running time n tends to ∞, is constructed. The chain is then run for a fairly long time n (say, 10^4 or 10^5), and X_n is output, in the hope that the chain has come close to equilibrium by then. But this is often just a matter of faith, or perhaps some vague handwaving arguments.

This situation is clearly unsatisfactory, and a substantial amount of effort has in recent years been put into attempts at rectifying it. In this chapter and in Chapters 10–12, we shall take a look at two different approaches. The one we consider in this chapter is to try to overcome the more serious problem (B) by establishing useful bounds for convergence rates of Markov chains. In general, this remains a difficult open problem, but in a number of specific situations, very good results have been obtained.

To illustrate the type of convergence rate results that can be obtained, and one of the main proof techniques, we will in this chapter focus on one particular example where the MCMC chain has been successfully analyzed, namely the random q-colorings in Example 7.3.

A variety of different (but sometimes related) techniques for proving fast convergence to equilibrium of Markov chains have been developed, including eigenvalue bounds, path and flow arguments, various comparisons between different chains, and the concept of strong stationary duality; see Chapter 14 for some references concerning these techniques. Another important technique, that we touched upon already in Chapter 5, is the use of couplings, and that is the approach we shall take here.

Let us consider the q-coloring example. Fix a graph $G = (V, E)$ and an integer q, and recall that $\rho_{G,q}$ is the probability distribution on $\{1, \ldots, q\}^V$ which is uniform over all $\xi \in \{1, \ldots, q\}^V$ that are valid q-colorings, i.e., over all assignments of colors $1, \ldots, q$ to the vertices of G with the property that no two vertices sharing an edge have the same color. We consider the Gibbs sampler described in Example 7.3, with the modification that the vertex to be updated is chosen as in the systematic sweep Gibbs sampler defined in (45). This means that instead of picking a vertex at random uniformly from $V = \{v_1, \ldots, v_k\}$, we scan systematically through the vertex set by updating vertex v_1 at time 1, v_2 at time 2, $\ldots$, v_k at time k, v_1 again at time $k + 1$, and so on as in (45).

It is natural to phrase the question about convergence rates for this MCMC algorithm (or others) as follows: Given $\varepsilon > 0$ (such as for instance $\varepsilon = 0.01$), how many iterations n of the algorithm do we need in order to make the total variation distance $d_{TV}(\mu^{(n)}, \rho_{G,q})$ less than ε? Here $\mu^{(n)}$ is the distribution of the chain after n iterations.

Theorem 8.1 *Let $G = (V, E)$ be a graph. Let k be the number of vertices in G, and suppose that any vertex $v \in V$ has at most d neighbors. Suppose furthermore that $q > 2d^2$. Then, for any fixed $\varepsilon > 0$, the number of iterations needed for the systematic sweep Gibbs sampler described above (starting from any fixed q-coloring ξ) to come within total variation distance ε of the target distribution $\rho_{G,q}$ is at most*

$$k \left(\frac{\log(k) + \log(\varepsilon^{-1}) - \log(d)}{\log \left(\frac{q}{2d^2} \right)} + 1 \right). \tag{49}$$

Before going to the proof of this result, some comments are in order:

1. The most important aspect of the bound in (49) is that it is bounded by

$$Ck(\log(k) + \log(\varepsilon^{-1}))$$

   for some constant $C < \infty$ that does not depend on k or on ε. This means that the number of iterations needed to come within total variation distance ε from the target distribution $\rho_{G,q}$ does not grow terribly quickly as $k \to \infty$ or as $\varepsilon \to 0$. It is easy to see that *any* algorithm for generating random q-colorings must have a running time that grows at least linearly in k (because it takes about time k even to *print* the result). The extra factor $\log(k)$ that we get here is not a particularly serious slowdown.

2. Our bound $q > 2d^2$ for when we get fast convergence is a fairly crude estimate. In fact, Jerrum [J] showed, by means of a refined version of the proof below, that a convergence rate of the same order of magnitude as in Theorem 8.1 takes place as soon as $q > 2d$, and it is quite likely that this bound can be improved even further.

3. If G is part of the square lattice (such as, for example, the graph in Figure 7), then $d = 4$, so that Theorem 8.1 gives fast convergence of the MCMC algorithm for $q \geq 33$. Jerrum's better bound gives fast convergence for $q \geq 9$.

4. It may seem odd that we obtain fast convergence for large q only, as one might intuitively think that it would be more difficult to simulate the larger q gets, due to the fact that the number of q-colorings on G is increasing in q. This is, however, misleading, and the correct intuition to have is instead the following. The larger q gets, the less dependent does the coloring of a vertex v become on its neighbors. If q is very large, we might pick the color at v uniformly at random, and have very little risk that this color is already taken by one of its neighbors. Hence, the difference between $\rho_{G,q}$ and uniform distribution over *all* elements of $\{1, \ldots, q\}^V$ becomes very

small in the limit as $q \to \infty$, and the latter distribution is of course easy to simulate: just assign i.i.d. colors (uniformly from $\{1, \ldots, q\}$) to the vertices.

Enough chat for now – it is time to do the five-page proof of the convergence rate result!

Proof of Theorem 8.1 As in the proof of Theorem 5.2, we will use a coupling argument: We shall run two $\{1, \ldots, q\}^V$-valued Markov chains $(X_0, X_1, \ldots)$ and $(X_0', X_1', \ldots)$ simultaneously. They will have the same transition matrices (namely, the ones given by the systematic sweep Gibbs sampler for random q-colorings of G, as described above). The difference will be that the first chain is started in the fixed state $X_0 = \xi$, whereas the second is started in a random state X_0' chosen according to the stationary distribution $\rho_{G,q}$. Then X_n' has distribution $\rho_{G,q}$ for all n, by the definition of stationarity. Also write $\mu^{(n)}$ for the distribution of the first chain $(X_0, X_1, \ldots)$ at time n; this is the chain that we are primarily interested in. We want to bound the total variation distance $d_{TV}(\mu^{(n)}, \rho_{G,q})$ between $\mu^{(n)}$ and the stationary distribution, and we shall see that $d_{TV}(\mu^{(n)}, \rho_{G,q})$ is close to 0 if $\mathbf{P}(X_n = X_n')$ is close to 1.

Recall from Example 7.3 that whenever a vertex v is chosen to be updated, we should pick a new color for v according to the uniform distribution on the set of colors that are not attained by any neighbor of v. One way to implement this concretely is to pick a random permutation

$$R = (R^1, \ldots, R^q)$$

of the set $\{1, \ldots, q\}$, chosen uniformly from the $q!$ different possible permutations (this is fairly easy; see Problem 8.1) and then let v get the first color of the permutation that is not attained by any neighbor of v.

Of course we need to pick a new (and independent) permutation at each update of a chain. However, nothing prevents us from using the same permutations for the chain $(X_0', X_1', \ldots)$ as for $(X_0, X_1, \ldots)$, and this is indeed what we shall do. Let $R_0, R_1, \ldots$ be an i.i.d. sequence of random permutations, each of them uniformly distributed on the set of permutations of $\{1, \ldots, q\}$. At each time n, the updates of the two chains use the permutation

$$R_n = (R_n^1, \ldots, R_n^q),$$

and the vertex v to be updated is assigned the new value

$$X_{n+1}(v) = R_n^i$$

where

$$i = \min\{j : X_n(w) \neq R_n^j \text{ for all neighbors } w \text{ of } v\}$$

in the first chain. In the second chain, we similarly set

$$X'_{n+1}(v) = R_n^{i'}$$

where

$$i' = \min\{j' : X'_n(w) \neq R_n^{j'} \text{ for all neighbors } w \text{ of } v\}.$$

This defines our coupling of $(X_0, X_1, \ldots)$ and $(X'_0, X'_1, \ldots)$. What we hope for is to have $X_T = X'_T$ at some (random, but not too large) time T, in which case we will also have $X_n = X'_n$ for all $n \geq T$ (because the coupling is defined in such a way that once the two chains coincide, they stay together forever). In order to estimate the probability that the configurations X_n and X'_n agree, let us first consider the probability that they agree *at a particular vertex*, i.e., that $X_n(v) = X'_n(v)$ for a given vertex v.

Consider the update of the two chains at a vertex v at time n, where we take $n \leq k$, so that in other words we are in the first sweep of the Gibbs sampler through the vertex set. We call the update **successful** if it results in having $X_{n+1}(v) = X'_{n+1}(v)$; otherwise we say that the update is **failed**. The probability of a successful update depends on the number of colors that are attained in the neighborhood of v in both configurations X_n and X'_n, and on the number of colors that are attained in each of them. Define

B_2 = the number of colors $r \in \{1, \ldots, q\}$ that are attained in the neighborhood of v in *both* X_n and X'_n,

B_1 = the number of colors $r \in \{1, \ldots, q\}$ that are attained in the neighborhood of v in *exactly one* of X_n and X'_n,

and

B_0 = the number of colors $r \in \{1, \ldots, q\}$ that are attained in the neighborhood of v in *neither* of X_n and X'_n,

and note that $B_0 + B_1 + B_2 = q$. Note also that if the first color R_n^1 in the permutation R_n is among the B_2 colors attained in the neighborhood of v in both configurations, then the Gibbs samplers just discard R_n^1 and look at R_n^2 instead, and so on. Therefore, the update is successful if and only if the first color in R_n that is attained in the neighborhood of v in *neither* of X_n and X'_n appears earlier in the permutation than the first color that is attained in the neighborhood of v in *exactly one* of X_n and X'_n. This event (of having a successful update) therefore has probability

$$\frac{B_0}{B_0 + B_1}$$

conditional on B_0, B_1 and B_2. In other words, we have[27]

$$\mathbf{P}(\text{failed update}) = \frac{B_1}{B_0 + B_1}. \tag{50}$$

We go on to estimate the right-hand side in (50). Clearly, $0 \le B_2 \le d$. Furthermore,

$$B_1 \le 2d - 2B_2, \tag{51}$$

because counting the neighbors in both configurations, there are in all at most $2d$ of them, and each color contributing to B_2 uses up two of them. We get

$$
\begin{aligned}
\mathbf{P}(\text{failed update}) \quad &= \quad \frac{B_1}{B_0 + B_1} = \frac{B_1}{q - B_2} \\
&\le \quad \frac{2d - 2B_2}{q - B_2} \le \frac{2d - B_2}{q - B_2} \\
&= \quad \frac{2d\left(1 - \frac{B_2}{2d}\right)}{q\left(1 - \frac{B_2}{q}\right)} \le \frac{2d}{q}
\end{aligned}
\tag{52}
$$

where the first inequality is just (51), while the final inequality is due to the assumption $q > 2d^2$, which implies $q > 2d$, which in turn implies $\left(1 - \frac{B_2}{q}\right) \ge \left(1 - \frac{B_2}{2d}\right)$.

Hence, we have, after k steps of the Markov chains (i.e., after the first sweep of the Gibbs samplers through the vertex set), that, for each vertex v,

$$\mathbf{P}(X_k(v) \ne X'_k(v)) \le \frac{2d}{q}. \tag{53}$$

Now consider updates during the second sweep of the Gibbs sampler, i.e., between times k and $2k$. For an update at time n during the second sweep to fail, the configurations X_n and X'_n need to differ in at least one neighbor of v. Each neighbor w has $X_n(w) \ne X'_n(w)$ with probability at most $\frac{2d}{q}$ (due to (53)), and summing over the at most d neighbors, we get that

$$\mathbf{P}(\text{discrepancy}) \le \frac{2d^2}{q} \tag{54}$$

where "discrepancy" is short for the event that there exists a neighbor w of v with $X_n(w) \ne X'_n(w)$. Given the event in (54), we have, by repeating the arguments in (50) and (52), that the conditional probability

---

[27] Our notation here is a bit sloppy, since it is really a conditional probability we are dealing with, because we are conditioning on B_0, B_1 and B_2.

P(failed update | discrepancy) of a failed update is bounded by $\frac{2d}{q}$. Hence,

$$
\begin{aligned}
P(\text{failed update}) \;&=\; P(\text{discrepancy})P(\text{failed update} \mid \text{discrepancy}) \\
&\leq\; \frac{4d^3}{q^2} = \frac{2d}{q}\left(\frac{2d^2}{q}\right).
\end{aligned}
$$

Hence, after $2k$ steps of the Markov chains, each vertex $v \in V$ has different colors in the two chains with probability at most

$$
P(X_{2k}(v) \neq X'_{2k}(v)) \leq \frac{2d}{q}\left(\frac{2d^2}{q}\right).
$$

By arguing in the same way for the third sweep as for the second sweep, we get that

$$
P(X_{3k}(v) \neq X'_{3k}(v)) \leq \frac{2d}{q}\left(\frac{2d^2}{q}\right)^2,
$$

and continuing in the obvious way, we get for $m = 4, 5, \ldots$ that

$$
P(X_{mk}(v) \neq X'_{mk}(v)) \leq \frac{2d}{q}\left(\frac{2d^2}{q}\right)^{m-1}. \tag{55}
$$

After this analysis of the probability that X_{mk} and X'_{mk} differ *at a given vertex*, we next want to estimate the probability $P(X_{mk} \neq X'_{mk})$ that the first chain fails to have *exactly the same configuration* as the second chain, at time mk. Since the event $X_{mk} \neq X'_{mk}$ implies that $X_{mk}(v) \neq X'_{mk}(v)$ for at least one vertex $v \in V$, we have

$$
\begin{aligned}
P(X_{mk} \neq X'_{mk}) \;&\leq\; \sum_{v \in V} P(X_{mk}(v) \neq X'_{mk}(v)) \\
&\leq\; k\frac{2d}{q}\left(\frac{2d^2}{q}\right)^{m-1} \tag{56} \\
&=\; \frac{k}{d}\left(\frac{2d^2}{q}\right)^{m} \tag{57}
\end{aligned}
$$

where the inequality in (56) is due to (55) and the assumption that the graph has k vertices.

Now let $A \subseteq \{1, \ldots, q\}^V$ be any subset of $\{1, \ldots, q\}^V$. By (33) and Problem 5.1, we have that

$$
\begin{aligned}
d_{\text{TV}}(\mu^{(mk)}, \rho_{G,q}) \;&=\; \max_{A \subseteq \{1,\ldots,q\}^V} \left|\mu^{(mk)}(A) - \rho_{G,q}(A)\right| \\
&=\; \max_{A \subseteq \{1,\ldots,q\}^V} \left|P(X_{mk} \in A) - P(X'_{mk} \in A)\right|. \tag{58}
\end{aligned}
$$

For any such A, we have

$$
\begin{aligned}
\mathbf{P}(X_{mk} &\in A) - \mathbf{P}(X'_{mk} \in A) \\
&= \quad \mathbf{P}(X_{mk} \in A, X'_{mk} \in A) + \mathbf{P}(X_{mk} \in A, X'_{mk} \notin A) \\
&\quad - \big(\mathbf{P}(X'_{mk} \in A, X_{mk} \in A) + \mathbf{P}(X'_{mk} \in A, X_{mk} \notin A)\big) \\
&= \quad \mathbf{P}(X_{mk} \in A, X'_{mk} \notin A) - \mathbf{P}(X'_{mk} \in A, X_{mk} \notin A) \\
&\leq \quad \mathbf{P}(X_{mk} \in A, X'_{mk} \notin A) \\
&\leq \quad \mathbf{P}(X_{mk} \neq X'_{mk}) \\
&\leq \quad \frac{k}{d}\left(\frac{2d^2}{q}\right)^m
\end{aligned}
\tag{59}
$$

where the last inequality uses (57). Similarly, we get

$$
\mathbf{P}(X'_{mk} \in A) - \mathbf{P}(X_{mk} \in A) \leq \frac{k}{d}\left(\frac{2d^2}{q}\right)^m .
\tag{60}
$$

Combining (59) and (60), we obtain

$$
\big|\mathbf{P}(X_{mk} \in A) - \mathbf{P}(X'_{mk} \in A)\big| \leq \frac{k}{d}\left(\frac{2d^2}{q}\right)^m .
\tag{61}
$$

By taking the maximum over all $A \subseteq \{1, \ldots, q\}^V$, and inserting into (58), we get that

$$
d_{\mathrm{TV}}(\mu^{(mk)}, \rho_{G,q}) \leq \frac{k}{d}\left(\frac{2d^2}{q}\right)^m ,
\tag{62}
$$

which tends to 0 as $m \to \infty$. Having established this bound, our next and final issue is:

How large does m need to be taken in order to make the right-hand side of (62) less than ε?

By setting

$$
\frac{k}{d}\left(\frac{2d^2}{q}\right)^m = \varepsilon
$$

and solving for m, we find that

$$
m = \frac{\log(k) + \log(\varepsilon^{-1}) - \log(d)}{\log\left(\frac{q}{2d^2}\right)}
$$

so that running the Gibbs sampler long enough to get at least this many scans through the vertex set gives $d_{TV}(\mu^{(mk)}, \rho_{G,q}) \leq \varepsilon$. To go from the number of scans m to the number of steps n of the Markov chain, we have to multiply by k, giving that

$$n = \frac{k(\log(k) + \log(\varepsilon^{-1}) - \log(d))}{\log\left(\frac{q}{2d^2}\right)} \tag{63}$$

should be enough. However, by taking n as in (63), we do not necessarily get an integer value for $m = \frac{n}{k}$, so to be on the safe side we should take n to be at least the smallest number which is greater than the right-hand side of (63) and which makes $\frac{n}{k}$ an integer. This means increasing n by at most k compared to (63), so that our final answer is that taking

$$n = k\left(\frac{\log(k) + \log(\varepsilon^{-1}) - \log(d)}{\log\left(\frac{q}{2d^2}\right)} + 1\right)$$

suffices, and Theorem 8.1 is (at last!) established. $\square$

Problems

8.1 **(4)** Describe a simple and efficient way to generate a random (uniform distribution) permutation of the set $\{1, \ldots, q\}$.

8.2 **(6) Bounding total variation distance using coupling.** Let π_1 and π_2 be probability distributions on some finite set S. Suppose that we can construct two random variables Y_1 and Y_2 such that

   (i) Y_1 has distribution π_1,
   (ii) Y_2 has distribution π_2, and
   (iii) $\mathbf{P}(Y_1 \neq Y_2) \leq \varepsilon$,

   for some given $\varepsilon \in [0, 1]$. Show that the total variation distance $d_{TV}(\pi_1, \pi_2)$ is at most ε. Hint: argue as in equations (59), (60) and (61) in the proof of Theorem 8.1.

8.3 **(8)** Explain where and why the assumption that $q > 2d^2$ is needed in the proof of Theorem 8.1.

8.4 **(10) Fast convergence for the random site Gibbs sampler.** Consider (instead of the systematic scan Gibbs sampler) the random site Gibbs sampler for random q-colorings, as in Example 7.3. Suppose that the graph $G = (V, E)$ has k vertices, and each vertex has at most d neighbors. Also suppose that $q > 2d^2$.

   (a) Show that for any given $v \in V$, the probability that v is chosen to be updated at some step during the first k iterations of the Markov chain is at least $1 - e^{-1}$. (Here $e \approx 2.7183$ is, of course, the base for the natural logarithm.)

   (b) Suppose that we run two copies of this Gibbs sampler, one starting in a fixed configuration, and one in equilibrium, similarly as in the proof of Theorem 8.1. Show that the coupling can be carried out in such a way that for any $v \in V$ and any m, the probability that the two chains have different colors at the vertex v

at time mk is at most

$$\left(e^{-1} + (1 - e^{-1})\frac{2d}{q}\right)\left(e^{-1} + (1 - e^{-1})\frac{2d^2}{q}\right)^{m-1}.$$

(c) Use the result in (b) to prove an analogue of Theorem 8.1 for the random site Gibbs sampler.

9

Approximate counting

Combinatorics is the branch of mathematics which deals with finite objects or sets, and the ways in which these can be combined. Basic objects that often arise in combinatorics are, e.g., graphs and permutations. Much of combinatorics deals with the following sort of problem:

Given some set S, what is the number of elements of S?

Let us give a few examples of such **counting problems**; the reader will probably be able to think of several interesting variations of these.

Example 9.1 What is the number of permutations $r = (r^1, \ldots, r^q)$ of the set $\{1, \ldots, q\}$ with the property that no two numbers that differ by exactly 1 are adjacent in the permutation?

Example 9.2 Imagine a chessboard, and a set of 32 domino tiles, such that one tile is exactly large enough to cover two adjacent squares of the chessboard. In how many different ways can the 32 tiles be arranged so as to cover the entire chessboard?

Example 9.3 Given a graph $G = (V, E)$, in how many ways can we pick a subset W of the vertex set V, with the property that no two vertices in W are adjacent in G? In other words, how many different feasible configurations exist for the hard-core model (see Example 7.1) on G?

Example 9.4 Given an integer q and a graph $G = (V, E)$, how many different q-colorings (Example 7.3) are there for G?

In this chapter, we are interested in algorithms for solving counting problems. For the purpose of illustrating some general techniques, we shall focus on the one in Example 9.4: the number of q-colorings of a graph. In particular, we shall see how (perhaps surprisingly!) the MCMC technique turns out to be useful in the context of counting problems. The same general approach has proved to be successful in a host of other counting problems, including counting of feasible hard-core configurations and of domino tilings, and estimation of the normalizing constant $Z_{G,\beta}$ in the so-called Ising model (which will be discussed in Chapter 11); see Sinclair [Si] for an overview.

The following algorithm springs immediately to mind as a solution to the problem of counting q-colorings.

> **Example 9.5: A naive algorithm for counting q-colorings.** If there were no restriction on the colorings, i.e., if all configurations in $\{1, \ldots, q\}^V$ were allowed, then the counting problem would be trivial: there are q^k of them, where k is the number of vertices in the graph. Moreover, it is trivial to make a list of all such configurations, for instance in some lexicographic order. Given a configuration $\xi \in \{1, \ldots, q\}^V$, the problem of determining whether ξ is a proper q-coloring of G is yet another triviality.[28] Hence, the following algorithm will work:
>
> > Go through all configurations in $\{1, \ldots, q\}^V$ in lexicographic order, check for each of them whether it is a q-coloring of G, and count the number of times the answer was "yes".
>
> This algorithm will certainly give the right answer. However, when k is large, it will take a very long time to run the algorithm, since it has to work itself through the list of all q^k configurations. For instance, when $q = 5$ and $k = 50$, there are $5^{50} \approx 10^{34}$ configurations to go through, which is impossible in practice. Therefore, this algorithm will only be useful for rather small graphs.

The feature which makes the algorithm in Example 9.5 unattractive is that the running time grows exponentially in the size k of the graph. The challenge in this type of situation is therefore to find faster algorithms. In particular, one is interested in **polynomial time algorithms**, i.e., algorithms with the property that there exists a polynomial $p(k)$ in the size k of the problem,[29] such that the running time is bounded by $p(k)$ for any k and any instance of the problem of size k. This is the same (see Problem 9.1) as asking for algorithms with a running time bounded by Ck^α for some constants C and α.

A polynomial time algorithm which solves a counting problem is called a **polynomial time counting scheme** for the problem. Sometimes, however, such an algorithm is not available, and we have to settle for something less, namely to *approximate* (rather than calculate exactly) the number of elements

---

[28] We just need to check, for each edge $e \in E$, that the endvertices of e have different colors.

[29] Here we measure the size of the problem in the number of vertices in the graph. This is usually the most natural choice for problems involving graphs, although sometimes there is reason take the number of edges instead. In Example 9.1, it is natural to take q as the size of the problem, whereas in generalizations of Example 9.2 to "chessboards" of arbitrary size, the size of the problem may be measured in the number of squares of the chessboard (or in the number of domino tiles). Common to all these measures of size is that the number of elements in the set to be counted grows (at least) exponentially in the size of the problem, making algorithms like the one in Example 9.5 infeasible.

of the set. Suppose that we have an algorithm which, in addition to the instance
of the counting problem, also takes a number $\varepsilon > 0$ as input. Suppose
furthermore that the algorithm has the properties that

(i) it always outputs an answer between $(1 - \varepsilon)N$ and $(1 + \varepsilon)N$, where N is
   the true answer to the counting problem, and
(ii) for any $\varepsilon > 0$, there exists a polynomial $p_\varepsilon(k)$ in the size k of the
   problem,[30] such that for any instance of size k, the algorithm terminates
   in at most $p_\varepsilon(k)$ steps.

We call such an algorithm a **polynomial time approximation scheme**. Given
a prespecified allowed relative error ε, the algorithm runs in polynomial time
in the size of the problem, and produces an answer which is within a multiplicative error ε of the true answer.

Sometimes, however, even this is too much to ask, and we have to be
content with an algorithm which produces an almost correct answer most of
the time, but which may produce a (vastly) incorrect answer with some positive
probability. More precisely, suppose that we have an algorithm which takes
$\varepsilon > 0$ and the instance of the counting problem as input, and has the properties
that

(i) with probability at least $\frac{2}{3}$, it outputs an answer between $(1 - \varepsilon)N$ and
   $(1 + \varepsilon)N$, where N is the true answer to the counting problem, and
(ii) for any $\varepsilon > 0$, there exists a polynomial $p_\varepsilon(k)$ in the size k of the problem,
   such that for any instance of size k, the algorithm terminates in at most
   $p_\varepsilon(k)$ steps.

Such an algorithm is called a **randomized polynomial time approximation
scheme**, and it is to the construction of such a scheme (for the q-coloring
counting problem) that this chapter is devoted.

One may (and should!) ask at this stage what is so special about the number
$\frac{2}{3}$. The answer is that it is, in fact, not special at all, and that it could be replaced
by any number strictly between $\frac{1}{2}$ and 1. For instance, for any $\delta > 0$, if we
have a randomized polynomial time approximation scheme (with the above
definition), then it is not difficult to build further upon it to obtain a randomized
polynomial time approximation scheme with the additional property that it
outputs an answer within a multiplicative error ε of the true answer, with
probability at least $1 - \delta$. We can thus get an answer within relative error
at most ε of the true answer, with probability as close to 1 as we may wish.
This property will be proved Problem 9.3 below.

---

[30] We allow $p_\varepsilon(k)$ to depend on ε in arbitrary fashion. Sometimes (although we shall not go into
this subtlety) there is reason to be restrictive about how fast $p_\varepsilon(k)$ may grow as $\varepsilon \to 0$.

Here is our main result concerning randomized polynomial time approximation schemes for random q-colorings.

Theorem 9.1 *Fix integers q and $d \geq 2$ such that $q > 2d^2$, and consider the problem of counting q-colorings for graphs in which each vertex has at most d neighbors. There exists a randomized polynomial time approximation scheme for this problem.*

Before going on with the proof of this result, some remarks are worth making:

1. Of course, an existence result (i.e., a statement of the form "there exists an algorithm such that . . .") of this kind is rather useless without an explicit description of the algorithm. Such a description, will, however, appear below as part of the proof.

2. The requirement that $q > 2d^2$ comes from Theorem 8.1, which will be used in the proof of Theorem 9.1. If we instead use Jerrum's better result mentioned in Remark 2 after Theorem 8.1, then we obtain the same result as in Theorem 9.1 whenever $q > 2d$.

3. The restriction to $d \geq 2$ is not a particularly severe one, since graphs with $d = 1$ consist just of

   (i) isolated vertices (having no neighbors), and

   (ii) pairs of vertices linked to each other by an edge but with no edge leading anywhere else,

   and the number of q-colorings of such graphs can be calculated directly (see Problem 9.2).

Another thing which it is instructive to do before the proof of Theorem 9.1 is to have a look at the following simple-minded attempt at a randomized algorithm, and to figure out why it does not work well in practice.

> **Example 9.6: Another naive algorithm for counting q-colorings.** Assume that $G = (V, E)$ is connected with k vertices, and write $Z_{G,q}$ for the number of q-colorings of G. Suppose that we assign each vertex independently a color from $\{1, \ldots, q\}$ chosen according to the uniform distribution, without regard to whether or not adjacent vertices have the same color. Then each configuration $\xi \in \{1, \ldots, q\}^V$ arises with the same probability $\frac{1}{q^k}$. Out of these q^k possible configurations, there are $Z_{G,q}$ that are q-colorings, whence the probability that this procedure yields a q-coloring is
>
> $$\frac{Z_{G,q}}{q^k}. \tag{64}$$

Let us now repeat this experiment n times, and write Y_n for the number of times that we got q-colorings. We clearly have

$$E[Y_n] = \frac{nZ_{G,q}}{q^k}$$

so that

$$E\left[\frac{q^k Y_n}{n}\right] = Z_{G,q}$$

which suggests that $\frac{q^k Y_n}{n}$ might be a good estimator of $Z_{G,q}$. Indeed, the Law of Large Numbers (Theorem 1.2) may be applied to show, for any $\varepsilon > 0$, that $\frac{q^k Y_n}{n}$ is between $(1 - \varepsilon)Z_{G,q}$ and $(1 + \varepsilon)Z_{G,q}$ with a probability which tends to 1 as $n \to \infty$.

But how large does n need to be? Clearly, $\frac{q^k Y_n}{n}$ is a very bad estimate as long as $Y_n = 0$, so we certainly need to pick n sufficiently large to make $Y_n > 0$ with a reasonably high probability. Unfortunately, this means that n has to be taken *very* large, as the following argument shows. In each simulation, we get a q-coloring with probability at most $\left(\frac{q-1}{q}\right)^{k-1}$; this is Problem 9.4. Hence,

$$
\begin{aligned}
P(Y_n > 0) \;&=\; P(\text{at least one of the first } n \text{ simulations yields a } q\text{-coloring}) \\
&\leq\; \sum_{i=1}^{n} P(\text{the } i^{\text{th}} \text{ simulation yields a } q\text{-coloring}) \\
&\leq\; n\left(\frac{q-1}{q}\right)^{k-1}.
\end{aligned}
$$

To make this probability reasonably large, say greater than $\frac{1}{2}$, we need to take $n \geq \frac{1}{2}\left(\frac{q}{q-1}\right)^{k-1}$. This quantity grows exponentially in k, making the algorithm useless for large graphs.

Let us pause for a moment and think about precisely what it is that makes the algorithm in Example 9.6 so creepingly slow. The reason is a combination of two facts: First, the probability in (64) that we are trying to estimate is extremely small: exponentially small in the number of vertices k. Second, to estimate a very small probability by means of simulation requires a very large number of simulations. In the algorithm that we are about to present as part of the proof of Theorem 9.1, one of the key ideas is to find other relevant probabilities to estimate, which have a more reasonable order of magnitude. Let us now turn to the proof.

First part of the proof of Theorem 9.1: a general description of the algorithm Suppose that the graph $G = (V, E)$ has k vertices and $\tilde{k}$ edges; by the assumption of the theorem we have that $\tilde{k} \leq dk$. Enumerate the edge set E as $\{e_1, \ldots, e_{\tilde{k}}\}$, and define the subgraphs $G_0, G_1, \ldots, G_{\tilde{k}}$ as follows. Let $G_0 = (V, \emptyset)$ be the graph with vertex set V and no edges, and for $j = 1, \ldots, \tilde{k}$, let

$$G_j = (V, \{e_1, \ldots, e_j\}).$$

In other words, G_j is the graph obtained from G by deleting the edges $e_{j+1}, \ldots, e_{\tilde{k}}$.

Next, let, for $j = 0, \ldots, \tilde{k}$, the number of q-colorings for the graph G_j be denoted by Z_j. Since $G_{\tilde{k}} = G$, we have that the number we wish to compute (or approximate) is $Z_{\tilde{k}}$. This number can be rewritten as

$$Z_{\tilde{k}} = \frac{Z_{\tilde{k}}}{Z_{\tilde{k}-1}} \times \frac{Z_{\tilde{k}-1}}{Z_{\tilde{k}-2}} \times \cdots \times \frac{Z_2}{Z_1} \times \frac{Z_1}{Z_0} \times Z_0. \tag{65}$$

If we can estimate each factor in the telescoped product in (65) to within sufficient accuracy, then we can multiply these estimates to get a reasonably accurate estimate of $Z_{\tilde{k}}$.

Note first that the last factor Z_0 is trivial to calculate: since G_0 has no edges, any assignment of colors from $\{1, \ldots, q\}$ to the vertices is a valid q-coloring, and since G_0 has k vertices, we have

$$Z_0 = q^k.$$

Consider next one of the other factors $\frac{Z_j}{Z_{j-1}}$ in (65). Write x_j and y_j for the endvertices of the edge e_j which is in G_j but not in G_{j-1}. By definition, Z_j is the number of q-colorings of the graph G_j. But the q-colorings of G_j are exactly those configurations $\xi \in \{1, \ldots, q\}^V$ that are q-colorings of G_{j-1} and that in addition satisfy $\xi(x_j) \neq \xi(y_j)$. Hence the ratio $\frac{Z_j}{Z_{j-1}}$ is exactly the proportion of q-colorings ξ of G_j that satisfy $\xi(x_j) \neq \xi(y_j)$. This means that

$$\frac{Z_j}{Z_{j-1}} = \rho_{G_{j-1},q}(X(x_j) \neq X(y_j)), \tag{66}$$

i.e., $\frac{Z_j}{Z_{j-1}}$ equals the probability that a random coloring X of G_{j-1}, chosen according to the uniform distribution $\rho_{G_{j-1},q}$, satisfies $X(x_j) \neq X(y_j)$.

The key point now is that the probability $\rho_{G_j,q}(X(x_j) \neq X(y_j))$ in (66) can be estimated using the simulation algorithm for $\rho_{G_{j-1},q}$ considered in Theorem 8.1. Namely, if we simulate a random q-coloring $X \in \{1, \ldots, q\}^V$ for G_{j-1} several times (using sufficiently many steps in the Gibbs sampler of Chapter 8), then the proportion of the simulations that result in a configuration with different colors at x_j and y_j is very likely to be close[31] to the desired

---

[31] This closeness is due to the Law of Large Numbers (Theorem 1.2).

expression in (66). We use this procedure to estimate each factor in the telescoped product in (65), and then multiply these to get a good estimate of the $Z_{\tilde{k}}$. $\qquad\square$

The second part of the proof of Theorem 9.1 consists of figuring out how many simulations we need to do for estimating each factor $\frac{Z_j}{Z_{j-1}}$, and how many steps of the Gibbs sampler we need to run in each simulation. For that, we first need three little lemmas:

Lemma 9.1 *Fix* $\varepsilon \in [0, 1]$, *let* k *be a positive integer, and let* $a_1, \ldots, a_k$ *and* $b_1, \ldots, b_k$ *be positive numbers satisfying*

$$\left(1 - \frac{\varepsilon}{2k}\right) \leq \frac{a_j}{b_j} \leq \left(1 + \frac{\varepsilon}{2k}\right)$$

for $j = 1, \ldots, k$. *Define the products* $a = \prod_{j=1}^{k} a_j$ *and* $b = \prod_{j=1}^{k} b_j$. *We then have*

$$1 - \varepsilon \leq \frac{a}{b} \leq 1 + \varepsilon. \tag{67}$$

Proof To prove the first inequality in (67), note that $(1 - \frac{\varepsilon}{2k})^2 \geq 1 - \frac{2\varepsilon}{2k}$, that $(1 - \frac{\varepsilon}{2k})^3 \geq (1 - \frac{\varepsilon}{2k})(1 - \frac{2\varepsilon}{2k}) \geq 1 - \frac{3\varepsilon}{2k}$, and so on, so that

$$\left(1 - \frac{\varepsilon}{2k}\right)^k \geq 1 - \frac{k\varepsilon}{2k}.$$

Hence,

$$\frac{a}{b} = \prod_{j=1}^{k} \frac{a_j}{b_j} \geq \prod_{j=1}^{k} \left(1 - \frac{\varepsilon}{2k}\right) = \left(1 - \frac{\varepsilon}{2k}\right)^k$$

$$\geq 1 - \frac{k\varepsilon}{2k} = 1 - \frac{\varepsilon}{2} \geq 1 - \varepsilon.$$

For the second inequality, we note that $e^{x/2} \leq 1 + x$ for all $x \in [0, 1]$ (plot the functions to see this!), so that

$$\frac{a}{b} = \prod_{j=1}^{k} \frac{a_j}{b_j} \leq \prod_{j=1}^{k} \left(1 + \frac{\varepsilon}{2k}\right) \leq \prod_{j=1}^{k} \exp\left(\frac{\varepsilon}{2k}\right)$$

$$= \exp\left(\frac{\varepsilon}{2}\right) \leq 1 + \varepsilon.$$

$\qquad\square$

Lemma 9.2 *Fix* $d \geq 2$ *and* $q > 2d^2$. *Let* $G = (V, E)$ *be a graph in which no vertex has more than* d *neighbors, and pick a random* q-*coloring* X *for* G

according to the uniform distribution $\rho_{G,q}$. *Then, for any two distinct vertices* $x, y \in V$, *the probability that* $X(x) \neq X(y)$ *satisfies*

$$\rho_{G,q}(X(x) \neq X(y)) \geq \frac{1}{2}. \tag{68}$$

Proof Note first that when x and y are neighbors in G, then (68) is trivial, since its left-hand side equals 1. We go on to consider the case where x and y are not neighbors.

Consider the following experiment, which is just a way of finding out the random coloring $X \in \{1, \ldots, q\}^V$: first look at the coloring $X(V \setminus \{x\})$ of all vertices except x, and only then look at the color at x. Because $\rho_{G,q}$ is uniform over all colorings, we have that the conditional distribution of the color $X(x)$ given $X(V \setminus \{x\})$ is uniform over all colors that are not attained by any neighbor of x. Clearly, x has at least $q - d$ colors to choose from, so the conditional probability of getting precisely the color that the vertex y got is at most $\frac{1}{q-d}$, regardless of what particular coloring $X(V \setminus \{x\})$ we got at the other vertices. It follows that $\rho_{G,q}(X(x) = X(y)) \leq \frac{1}{q-d}$, so that

$$
\begin{aligned}
\rho_{G,q}(X(x) \neq X(y)) &= 1 - \rho_{G,q}(X(x) = X(y)) \\
&\geq 1 - \frac{1}{q-d} \geq 1 - \frac{1}{2d^2 - d} \\
&\geq 1 - \frac{1}{2} = \frac{1}{2}.
\end{aligned}
$$

$\square$

Lemma 9.3 *Fix* $p \in [0, 1]$ *and a positive integer* n. *Toss a coin with heads-probability* p *independently* n *times, and let* H *be the number of heads. Then, for any* $a > 0$, *we have*

$$\mathbf{P}(|H - np| \geq a) \leq \frac{n}{4a^2}.$$

Proof Note that H is a binomial (n, p) random variable; see Example 1.3. Therefore it has mean $\mathbf{E}[H] = np$ and variance $\mathbf{Var}[H] = np(1 - p)$. Hence, Chebyshev's inequality (Theorem 1.1) gives

$$
\begin{aligned}
\mathbf{P}(|H - np| \geq a) &\leq \frac{np(1 - p)}{a^2} \\
&\leq \frac{n}{4a^2}
\end{aligned}
$$

using the fact that $p(1 - p) \leq \frac{1}{4}$ for all $p \in [0, 1]$. $\square$

Second part of the proof of Theorem 9.1 We need some notation. For $j = 1, \ldots, \tilde{k}$, write Y_j for the algorithm's (random) estimator of $\frac{Z_j}{Z_{j-1}}$. Also define the products $Y = \prod_{j=1}^{\tilde{k}} Y_j$ and

$$Y^* = Z_0 Y = q^k Y = q^k \prod_{j=1}^{\tilde{k}} Y_j \,. \tag{69}$$

Because of (65), we take Y^* as the estimator of the desired quantity $Z_{\tilde{k}}$, i.e., as the output of the algorithm. First, however, we need to generate, for $j = 1, \ldots, \tilde{k}$, the estimator Y_j of $\frac{Z_j}{Z_{j-1}}$. How much error can we allow in each of these estimators $Y_1, \ldots, Y_{\tilde{k}}$? Well, suppose that we make sure that

$$\left(1 - \frac{\varepsilon}{2\tilde{k}}\right) \frac{Z_j}{Z_{j-1}} \leq Y_j \leq \left(1 + \frac{\varepsilon}{2\tilde{k}}\right) \frac{Z_j}{Z_{j-1}} \tag{70}$$

for each j. This is the same as

$$1 - \frac{\varepsilon}{2\tilde{k}} \leq \frac{Y_j}{Z_j/Z_{j-1}} \leq 1 + \frac{\varepsilon}{2\tilde{k}} \,,$$

and Lemma 9.1 therefore guarantees that

$$1 - \varepsilon \leq \frac{Y}{\prod_{j=1}^{\tilde{k}}(Z_j/Z_{j-1})} \leq 1 - \varepsilon \,,$$

which is the same as

$$1 - \varepsilon \leq \frac{Y}{Z_{\tilde{k}}/Z_0} \leq 1 - \varepsilon \,.$$

The definition (69) of our estimator Y^* gives

$$1 - \varepsilon \leq \frac{Y^*}{Z_{\tilde{k}}} \leq 1 + \varepsilon$$

which we can rewrite as

$$(1 - \varepsilon) Z_{\tilde{k}} \leq Y^* \leq (1 + \varepsilon) Z_{\tilde{k}} \,. \tag{71}$$

This is exactly what we need. It therefore only remains to obtain Y_j's that satisfy (70). We can rewrite (70) as

$$-\frac{\varepsilon}{2\tilde{k}} \frac{Z_j}{Z_{j-1}} \leq Y_j - \frac{Z_j}{Z_{j-1}} \leq \frac{\varepsilon}{2\tilde{k}} \frac{Z_j}{Z_{j-1}} \,. \tag{72}$$

Due to (66) and Lemma 9.2, we have that $\frac{Z_j}{Z_{j-1}} \geq \frac{1}{2}$. Hence, (72) and (70) follow if we can make sure that

$$-\frac{\varepsilon}{4\tilde{k}} \leq Y_j - \frac{Z_j}{Z_{j-1}} \leq \frac{\varepsilon}{4\tilde{k}} \,. \tag{73}$$

Recall that Y_j is obtained by simulating random q-colorings for G_{j-1} several times, by means of the Gibbs sampler in Chapter 8, and taking Y_j to be the proportion of the simulations that result in a q-coloring ξ satisfying $\xi(x_j) \neq \xi(y_j)$. There are two sources of error in this procedure, namely

 (i) the Gibbs sampler (which we start in some fixed but arbitrary q-coloring ξ) is only run for a finite number n of steps, so that the distribution $\mu^{(n)}$ of the coloring that it produces may differ somewhat from the target distribution $\rho_{G_j, q}$, and

 (ii) only finitely many simulations are done, so the proportion Y_j resulting in q-colorings ξ with $\xi(x_j) \neq \xi(y_j)$ may differ somewhat from the expected value $\mu^{(n)}(X(x_j) \neq X(y_j))$.

According to (73), Y_j is allowed to differ from $\frac{Z_j}{Z_{j-1}}$ (i.e., from $\rho_{G_{j-1}, q}(X(x_j) \neq X(y_j))$, by (66)) by at most $\frac{\varepsilon}{4\tilde{k}}$. One way to accomplish this is to make sure that

$$\left| \mu^{(n)}(X(x_j) \neq X(y_j)) - \rho_{G_{j-1}, q}(X(x_j) \neq X(y_j)) \right| \leq \frac{\varepsilon}{8\tilde{k}} \tag{74}$$

and that

$$\left| Y_j - \mu^{(n)}(X(x_j) \neq X(y_j)) \right| \leq \frac{\varepsilon}{8\tilde{k}}, \tag{75}$$

In other words, the leeway $\frac{\varepsilon}{4\tilde{k}}$ allowed by (66) is split up equally between the two error sources in (i) and (ii).

Let us first consider how many steps of the Gibbs sampler we need to run in order to make the error from (i) small enough so that (74) holds. By the formula (33) for total variation distance d_{TV}, it is enough to run the Gibbs sampler for a sufficiently long time n to make

$$d_{\mathrm{TV}}(\mu^{(n)}, \rho_{G_{j-1}, q}) \leq \frac{\varepsilon}{8\tilde{k}},$$

and Theorem 8.1 is exactly suited for determining such an n. Indeed, it suffices, by Theorem 8.1, to take

$$
\begin{aligned}
n &= k \left(\frac{\log(k) + \log(\frac{8\tilde{k}}{\varepsilon}) - \log(d)}{\log(\frac{q}{2d^2})} + 1 \right) \\
&\leq k \left(\frac{\log(k) + \log(\frac{8dk}{\varepsilon}) - \log(d)}{\log(\frac{q}{2d^2})} + 1 \right) \\
&= k \left(\frac{2\log(k) + \log(\varepsilon^{-1}) + \log(8)}{\log(\frac{q}{2d^2})} + 1 \right) \tag{76}
\end{aligned}
$$

where the inequality is because $\tilde{k} \leq dk$.

Next, we consider the number of simulations of q-colorings of G_{j-1} needed to make the error from (ii) small enough so that (75) holds, with sufficiently high probability. By part (i) of the definition of a randomized polynomial time approximation scheme, the algorithm is allowed to fail (i.e., to produce an answer Y^* that does not satisfy (71)) with probability at most $\frac{1}{3}$. Since there are $\tilde{k}$ estimators Y_j to compute, we can allow each one to fail (i.e., to disobey (75)) with probability $\frac{1}{3\tilde{k}}$. The probability that the algorithm fails is then at most $\tilde{k}\frac{1}{3\tilde{k}} = \frac{1}{3}$, as desired.

Suppose now that we make m simulations[32] when generating Y_j, and write H_j for the number of them that result in colorings ξ with $\xi(x_j) \neq \xi(y_j)$. Then

$$Y_j = \frac{H_j}{m}.$$

By multiplying both sides of (75) with m, we get that (75) is equivalent to

$$|H_j - mp| \leq \frac{\varepsilon m}{8\tilde{k}},$$

where p is defined by $p = \mu^{(n)}(X(x_j) \neq X(y_j))$. But the distribution of H_j is precisely the distribution of the number of heads when we toss m coins with heads-probability p. Lemma 9.3[33] therefore gives

$$\mathbf{P}\left[|H_j - mp| > \frac{\varepsilon m}{8\tilde{k}}\right] \leq \frac{m}{4\left(\frac{\varepsilon m}{8\tilde{k}}\right)^2}$$

$$= \frac{16\tilde{k}^2}{\varepsilon^2 m} \qquad (77)$$

and we need to make this probability less than $\frac{1}{3\tilde{k}}$. Setting the expression in (77) equal to $\frac{1}{3\tilde{k}}$ and solving for m gives

$$m = \frac{48\tilde{k}^3}{\varepsilon^2},$$

and this is the number of simulations we need to make for each Y_j. Using $\tilde{k} \leq dk$ again, we get

$$m \leq \frac{48d^3k^3}{\varepsilon^2}.$$

---

[32] Each time, we use the Gibbs sampler starting in the same fixed q-coloring, and run it for n steps, with n satisfying (76).

[33] An alternative to using Lemma 9.3 (and therefore indirectly Chebyshev's inequality), which leads to sharper upper bounds on how large m needs to be, is to use the so-called **Chernoff bound** for the binomial distribution; see, e.g., [MR].

Let us summarize: The algorithm has $\tilde{k}$ factors Y_j to compute. Each one is obtained using no more than $\frac{48d^3k^3}{\varepsilon^2}$ simulations, and, by (76), each simulation requires no more than $k\left(\frac{2\log(k)+\log(\varepsilon^{-1})+\log(8)}{\log\left(\frac{q}{2d^2}\right)}+1\right)$ steps of the Gibbs sampler. The total number of steps needed is therefore at most

$$dk \times \frac{48d^3k^3}{\varepsilon^2} \times k \left(\frac{2\log(k)+\log(\varepsilon^{-1})+\log(8)}{\log\left(\frac{q}{2d^2}\right)}+1\right)$$

which is of the order $Ck^5 \log(k)$ as $k \to \infty$ for some constant C that does not depend on k. This is less than Ck^6, so the total number of iterations in the Gibbs sampler grows no faster than polynomially. Since, clearly, the running times of all other parts of the algorithm are asymptotically negligible compared to these Gibbs sampler iterations, Theorem 9.1 is established. □

Problems

9.1 (3) Suppose that we have an algorithm whose running time is bounded by a polynomial $p(k)$, where k is the "size" (see Footnote 29 in this chapter) of the input. Show that there exist constants C and α such that the running time is bounded by Ck^α.

9.2 (3) Suppose that G is a graph consisting of k isolated vertices (i.e., vertices that are not the endpoint of any edge) plus l pairs of vertices where in each pair the two vertices are linked by an edge, but have no other neighbors. Show that the number of q-colorings of G is $q^{k+l}(q - 1)^l$.

9.3 (8) The definition of a randomized polynomial time approximation scheme allows the algorithm to produce, with probability $\frac{1}{3}$, an output which is incorrect, in the sense that it is *not* between $(1 - \varepsilon)N$ and $(1 + \varepsilon)N$, where N is the true answer to the counting problem. The error probability $\frac{1}{3}$ can be cut to any given $\delta > 0$ by the following method: Run the algorithm many (say m, where m is odd) times, and take the median of the outputs (i.e., the $\frac{m+1}{2}$th largest output). Show that this works for m large enough, and give an explicit bound (depending on δ) for determining how large "large enough" is.

9.4 (7) Let $G = (V, E)$ be a connected graph on k vertices, and pick $X \in \{1, \ldots, q\}^V$ at random, with probability $\frac{1}{q^k}$ for each configuration. Show that the probability that X is a q-coloring is at most

$$\left(\frac{q-1}{q}\right)^{k-1}.$$

Hint: enumerate the vertices $v_1, \ldots, v_k$ in such a way that each v_i has at least one edge to some earlier vertex. Then imagine revealing the colors of $v_1, v_2, \ldots$ one at a time, each time considering the conditional probability of not getting the same color as a neighboring vertex.

10

The Propp–Wilson algorithm

Recall, from the beginning of Chapter 8, the problems (A) and (B) with the MCMC method. In that chapter, we saw one approach to solving these problems, namely to prove that an MCMC chain converges sufficiently quickly to its equilibrium distribution.

In the early 1990's, some ideas about a radically different approach began to emerge. The breakthrough came in a 1996 paper by Jim Propp and David Wilson [PW], both working at MIT at that time, who presented a refinement of the MCMC method, yielding an algorithm which simultaneously solves problems (A) and (B) above, by

(A*) producing an output whose distribution is *exactly* the equilibrium distribution π, and
(B*) determining automatically when to stop, thus removing the need to compute any Markov chain convergence rates beforehand.

This algorithm has become known as the **Propp–Wilson algorithm**, and is the main topic of this chapter. The main feature distinguishing the Propp–Wilson algorithm from ordinary MCMC algorithms is that it involves running not only one Markov chain, but several copies of it,[34] with different initial values. Another feature which is important (we shall soon see why) is that the chains are not run from time 0 and onwards, but rather from some time in the (possibly distant) past, and up to time 0.

Due to property (A*) above, the Propp–Wilson algorithm is sometimes said to be an **exact**, or **perfect** simulation algorithm.

We go on with a more specific description of the algorithm. Suppose that we want to sample from a given probability distribution π on a finite set

---

[34] That is, we are working with a coupling of Markov chains; see Footnote 17 in Chapter 5. For reasons that will become apparent, Propp and Wilson called their algorithm **coupling from the past**.

$S = \{s_1, \ldots, s_k\}$. As in ordinary MCMC, we construct a reversible, irreducible and aperiodic Markov chain with state space S and stationary distribution π. Let P be the transition matrix of the chain, and let $\phi : S \times [0, 1] \to S$ be some valid update function, as defined in Chapter 3. Furthermore, let $N_1, N_2, \ldots$ be an increasing sequence of positive integers; a common and sensible[35] choice is to take $(N_1, N_2, \ldots) = (1, 2, 4, 8, \ldots)$. (The negative numbers $-N_1, -N_2, \ldots$ will be used as "starting times" for the Markov chains.) Finally, suppose that $U_0, U_{-1}, U_{-2}, \ldots$ is a sequence of i.i.d. random numbers, uniformly distributed on $[0, 1]$. The algorithm now runs as follows.

1. Set $m = 1$.
2. For each $s \in \{s_1, \ldots, s_k\}$, simulate the Markov chain starting at time $-N_m$ in state s, and running up to time 0 using update function ϕ and random numbers $U_{-N_m+1}, U_{-N_m+2}, \ldots, U_{-1}, U_0$ (these are the same for each of the k chains).
3. If all k chains in Step 2 end up in the same state s' at time 0, then output s' and stop. Otherwise continue with Step 4.
4. Increase m by 1, and continue with Step 2.

It is important that at the mth time that we come to Step 2, and need to use the random numbers $U_{-N_m+1}, U_{-N_m+2}, \ldots, U_{-1}, U_0$, that we actually *reuse* those random numbers $U_{-N_{m-1}+1}, U_{-N_{m-1}+2}, \ldots, U_{-1}, U_0$ that we have used before. This is necessary for the algorithm to work correctly (i.e., to produce an unbiased sample from π; see Example 10.2 below), but also somewhat cumbersome, since it means that we must store a (perhaps very long) sequence of random numbers, for possible further use.[36]

In Figure 8, we consider a simple example with $(N_1, N_2, \ldots) = (1, 2, 4, 8, \ldots)$ and state space $S = \{s_1, s_2, s_3\}$. Since $N_1 = 1$, we start by running the chain from time -1 to time 0. Suppose (as in the top part of Figure 8) that it turns out that

$$\begin{cases} \phi(s_1, U_0) = s_1 \\ \phi(s_2, U_0) = s_2 \\ \phi(s_3, U_0) = s_1 . \end{cases}$$

Hence the state at time 0 can take two different values (s_1 or s_2) depending on the state at time -1, and we therefore try again with starting time $-N_2 = -2$.

---

[35] See Problem 10.1.

[36] An ingenious way to circumvent this problem of having to store a long sequence of random numbers will be discussed in Chapter 12.

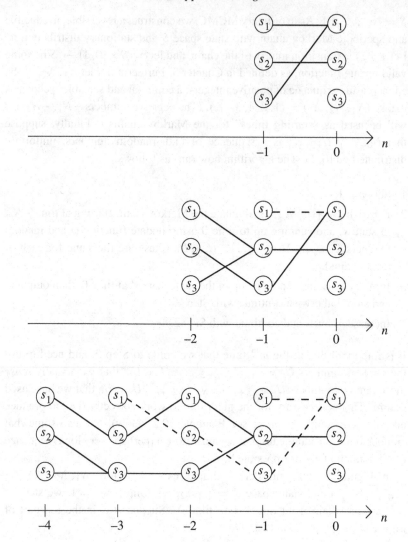

Fig. 8. A run of the Propp–Wilson algorithm with $N_1 = 1$, $N_2 = 2$, $N_3 = 4$, and state space $S = \{s_1, s_2, s_3\}$. Transitions that are carried out in the running of the algorithm are indicated with solid lines; others are dashed.

We then get

$$
\begin{cases}
\phi(\phi(s_1, U_{-1}), U_0) = \phi(s_2, U_0) = s_2 \\
\phi(\phi(s_2, U_{-1}), U_0) = \phi(s_3, U_0) = s_1 \\
\phi(\phi(s_3, U_{-1}), U_0) = \phi(s_2, U_0) = s_2
\end{cases}
$$

which again produces two different values at time 0. We are therefore again
forced to start the chains from an earlier starting time $-N_3 = -4$. This yields

$$\begin{cases} \phi(\phi(\phi(\phi(\phi(s_1, U_{-3}), U_{-2}), U_{-1}), U_0) = \cdots = s_2 \\ \phi(\phi(\phi(\phi(\phi(s_2, U_{-3}), U_{-2}), U_{-1}), U_0) = \cdots = s_2 \\ \phi(\phi(\phi(\phi(\phi(s_3, U_{-3}), U_{-2}), U_{-1}), U_0) = \cdots = s_2 \,. \end{cases}$$

This time, we get to state s_2 at time 0, regardless of the starting value at time
-4. The algorithm therefore stops with output equal to s_2. Note that if we were
to continue and run the chains starting at times -8, -16 and so on, then we
would keep getting the same output (state s_2) forever. Hence, the output can
be thought of as the value at time 0 of a chain that has been running since time
$-\infty$ (whatever that means!), and which therefore has reached equilibrium.
This is the intuition for why the Propp–Wilson algorithm works; this intuition
will be turned into mathematical rigor in the proof of Theorem 10.1 below.

Note that the Propp–Wilson algorithm contains a potentially unbounded
loop, and that we therefore don't have any general guarantee that the algorithm
will ever terminate. In fact, it may fail to terminate if the update function ϕ
is chosen badly; see Problem 10.2. On the other hand, it is often possible to
show that the algorithm terminates with probability 1.[37] In that case, it outputs
an unbiased sample from the desired distribution π, as stated in the following
theorem.

Theorem 10.1 *Let P be the transition matrix of an irreducible and aperiodic
Markov chain with state space $S = \{s_1, \ldots, s_k\}$ and stationary distribution
$\pi = (\pi_1, \ldots, \pi_k)$. Let ϕ be a valid update function for P, and consider
the Propp–Wilson algorithm as above with $(N_1, N_2, \ldots) = (1, 2, 4, 8, \ldots)$.
Suppose that the algorithm terminates with probability 1, and write Y for its
output. Then, for any $i \in \{1, \ldots, k\}$, we have*

$$\mathbf{P}(Y = s_i) = \pi_i \,. \tag{78}$$

Proof Fix an arbitrary state $s_i \in S$. In order to prove (78), it is enough to show
that for any $\varepsilon > 0$, we have

$$|\mathbf{P}(Y = s_i) - \pi_i| \le \varepsilon \,. \tag{79}$$

So fix an arbitrary $\varepsilon > 0$. By the assumption that the algorithm terminates with

---

[37] To show this, it helps to know that there is a so-called **0-1 law** for the termination of the
Propp–Wilson algorithm, meaning that the probability that it terminates must be either 0 or
1. Hence, it is enough to show that $\mathbf{P}(\text{algorithm terminates}) > 0$ in order to show that
$\mathbf{P}(\text{algorithm terminates}) = 1$.

probability 1, we can make sure that

$$\mathbf{P}(\text{the algorithm does not need to try starting times earlier than } -N_M) \quad (80)$$
$$\geq 1 - \varepsilon,$$

by picking M sufficiently large. Fix such an M, and imagine running a Markov chain from time $-N_M$ up to time 0, with the same update function ϕ and the same random numbers $U_{-N_M+1}, \ldots, U_0$ as in the algorithm, but *with the initial state at time* $-N_M$ *chosen according to the stationary distribution* π. Write $\tilde{Y}$ for the state at time 0 of this imaginary chain. Since π is stationary, we have that $\tilde{Y}$ has distribution π. Furthermore, $\tilde{Y} = Y$ if the event in (80) happens, so that

$$\mathbf{P}(Y \neq \tilde{Y}) \leq \varepsilon.$$

We therefore get

$$
\begin{aligned}
\mathbf{P}(Y = s_i) - \pi_i &= \mathbf{P}(Y = s_i) - \mathbf{P}(\tilde{Y} = s_i) \\
&\leq \mathbf{P}(Y = s_i, \tilde{Y} \neq s_i) \\
&\leq \mathbf{P}(Y \neq \tilde{Y}) \leq \varepsilon \quad (81)
\end{aligned}
$$

and similarly

$$
\begin{aligned}
\pi_i - \mathbf{P}(Y = s_i) &= \mathbf{P}(\tilde{Y} = s_i) - \mathbf{P}(Y = s_i) \\
&\leq \mathbf{P}(\tilde{Y} = s_i, Y \neq s_i) \\
&\leq \mathbf{P}(Y \neq \tilde{Y}) \leq \varepsilon. \quad (82)
\end{aligned}
$$

By combining (81) and (82), we obtain (79), as desired. □

At this stage, a very natural objection regarding the usefulness of the Propp–Wilson algorithm is the following: Suppose that the state space S is very large,[38] as, e.g., in the hard-core model example in Chapter 7. How on earth can we then run the chains from all possible starting values? This will simply take too much computer time to be doable in practice.

The answer is that various ingenious techniques have been developed for representing the chains in such a way that not all of the chains have to be simulated explicitly in order to keep track of their values. Amongst the most important such techniques is a kind of "sandwiching" idea which works for Markov chains that obey certain monotonicity properties; this will be the topic of the next chapter.

---

[38] Otherwise there is no need for a Propp–Wilson algorithm, because if the state space S is small, then the very simple simulation method in (42) can be used.

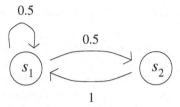

0.5

Fig. 9. Transition graph for the Markov chain used as counterexample to the modified algorithms in Examples 10.1 and 10.2.

Let us close the present chapter by discussing a couple of very tempting (but unfortunately incorrect) attempts at simplifying the Propp–Wilson algorithm. The fact that these close variants do not work might possibly explain why the Propp–Wilson algorithm was not discovered much earlier.

Example 10.1: "Coupling to the future". One of the most common reactions among bright students upon having understood the Propp–Wilson algorithm is the following.

> OK, that's nice. But why bother with all these starting times further and further into the past? Why not simply start chains in all possible states at time 0, and then run them forwards in time until the first time N at which they coalesce, and then output their common value?

This is indeed extremely tempting, but as it turns out, it gives biased samples in general. To see this, consider the following simple example. Let $(X_0, X_1, \ldots)$ be a Markov chain with state space $S = \{s_1, s_2\}$ and transition matrix

$$P = \begin{bmatrix} 0.5 & 0.5 \\ 1 & 0 \end{bmatrix}.$$

See the transition graph in Figure 9. Clearly, the chain is reversible with stationary distribution

$$\pi = (\pi_1, \pi_2) = \left(\frac{2}{3}, \frac{1}{3} \right). \tag{83}$$

Suppose that we run two copies of this chain starting at time 0, one in state s_1 and the other in state s_2. They will coalesce (take the same value) for the first time at some random time N. Consider the situation at time $N - 1$. By the definition of N, they cannot be in the same state at time $N - 1$. Hence one of the chains is in state s_2 at time $N - 1$. But the transition matrix tells us that this chain will with probability 1 be in state s_1 at the next instant, which is time N. Hence the chains are with probability 1 in state s_1 at the first time of coalescence, so that this modified Propp–Wilson algorithm outputs state s_1 with probability 1. This is not in agreement with the stationary distribution in (83), and hence the algorithm is incorrect.

Example 10.2 Here's another common suggestion for simplification of the Propp–Wilson algorithm:

> The need to reuse the random variables $U_{-N_m+1}, U_{-N_m+2}, \ldots, U_0$ when restarting the chains at time $-N_{m+1}$ is really annoying. Why don't we simply generate some new random numbers and use them instead?

As an example to show that this modification, like the one in Example 10.1, gives biased samples, we use again the Markov chain in Figure 9. Let us suppose that we take the Propp–Wilson algorithm for this chain, with $(N_1, N_2, \ldots) = (1, 2, 4, 8, \ldots)$ and update function ϕ given by (21), but modify it according to the suggested use of fresh new random numbers at each round. Let Y denote the output of this modified algorithm, and define the random variable M as the largest m for which the algorithm decides to simulate chains starting at time $-N_m$. A direct calculation gives

$$
\begin{aligned}
\mathbf{P}(Y = s_1) &= \sum_{m=1}^{\infty} \mathbf{P}(M = m, Y = s_1) \\
&\geq \mathbf{P}(M = 1, Y = s_1) + \mathbf{P}(M = 2, Y = s_1) \\
&= \mathbf{P}(M = 1)\mathbf{P}(Y = s_1 \mid M = 1) + \mathbf{P}(M = 2)\mathbf{P}(Y = s_1 \mid M = 2) \\
&= \frac{1}{2} \cdot 1 + \frac{3}{8} \cdot \frac{2}{3} \\
&= \frac{3}{4} > \frac{2}{3}
\end{aligned}
\tag{84}
$$

(of course, some details are omitted in line (84) of the calculation; see Problem 10.3). Hence, the distribution of the output Y does not agree with the distribution π given by (83). The proposed modified algorithm is therefore incorrect.

Problems

10.1 **(5)** For a given Propp–Wilson algorithm, define the integer-valued random variable N^* as

$$
N^* = \min\{n : \text{the chains starting at time } -n \text{ coalesce by time } 0\} \, .
$$

If we now choose starting times $(N_1, N_2, \ldots) = (1, 2, 3, 4, \ldots)$, then the total number of time units that we need to run the Markov chains is

$$
1 + 2 + 3 + \cdots + N^* = \frac{N^*(N^* + 1)}{2} \, ,
$$

which grows like the square of N^*. Show that if we instead use $(N_1, N_2, \ldots) = (1, 2, 4, 8, \ldots)$, then the total number of iterations executed is bounded by $4N^*$, so that in particular it grows only linearly in N^*, and therefore is much more efficient.

10.2 **(8) The choice of update function matters.** Recall from Problem 3.2 that for a given Markov chain, there may be more than one possible choice of valid update function. For ordinary MCMC simulation, this choice is more or less inconsequential, but for the Propp–Wilson algorithm, it is often extremely important.

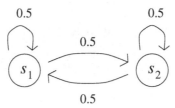

Fig. 10. Transition graph for the Markov chain considered in Problem 10.2.

Consider for instance the Markov chain[39] with state space $S = \{s_1, s_2\}$, transition matrix

$$P = \begin{bmatrix} 0.5 & 0.5 \\ 0.5 & 0.5 \end{bmatrix},$$

and transition graph as in Figure 10. Suppose that we run a Propp–Wilson algorithm for this Markov chain, with $(N_1, N_2, \ldots) = (1, 2, 4, 8, \ldots)$.

(a) One possible choice of valid update function is to set

$$\phi(s_i, x) = \begin{cases} s_1 & \text{for } x \in [0, 0.5) \\ s_2 & \text{for } x \in [0.5, 1] \end{cases}$$

for $i = 1, 2$. Show that with this choice of ϕ, the algorithm terminates (with probability 1) immediately after having run the chains from time $-N_1 = -1$ to time 0.

(b) Another possible choice of valid update function is to set

$$\phi(s_1, x) = \begin{cases} s_1 & \text{for } x \in [0, 0.5) \\ s_2 & \text{for } x \in [0.5, 1] \end{cases}$$

and

$$\phi(s_2, x) = \begin{cases} s_2 & \text{for } x \in [0, 0.5) \\ s_1 & \text{for } x \in [0.5, 1]. \end{cases}$$

Show that with this choice of ϕ, the algorithm *never* terminates.

10.3 (6) Verify that the calculation in equation (84) of Example 10.2 is correct.

---

[39] This particular Markov chain is even more trivial than most of the other examples that we have considered, because it produces an i.i.d. sequence of 0's and 1's. But that is beside the point of this problem.

11

Sandwiching

For the Propp–Wilson algorithm to be of any use in practice, we need to make it work also in cases where the state space S of the Markov chain is very large. If S contains k elements, then the Propp–Wilson algorithm involves running k different Markov chains in parallel, which is not doable in practice when k is very large. We therefore need to find some way to represent the Markov chains (or to use some other trick) that allows us to just keep track of a smaller set of chains.

In this chapter, we will take a look at **sandwiching**, which is the most famous (and possibly the most important) such idea for making the Propp–Wilson algorithm work on large state spaces. The sandwiching idea applies to Markov chains obeying certain monotonicity properties with respect to some ordering of the state space; several important examples fit into this context, but it is also important to keep in mind that there are many Markov chains for which sandwiching does *not* work.

To explain the idea, let us first consider a very simple case. Fix k, let the state space be $S = \{1, \ldots, k\}$, and let the transition matrix be given by

$$P_{11} = P_{12} = \frac{1}{2},$$

$$P_{kk} = P_{k,k-1} = \frac{1}{2},$$

and, for $i = 2, \ldots, k - 1$,

$$P_{i,i-1} = P_{i,i+1} = \frac{1}{2}.$$

All the other entries of the transition matrix are 0. In words, what the Markov chain does is that at each integer time it takes one step up or one step down the "ladder" $\{1, \ldots, k\}$, each with probability $\frac{1}{2}$; if the chain is already on top of the ladder (state k) and tries to take a step up, then it just stays where it is, and similarly at the bottom of the ladder. Let us call this Markov chain the **ladder walk** on k vertices. By arguing as in Example 6.2, it is not hard to show that this Markov chain has stationary distribution π given by

$$\pi_i = \frac{1}{k} \quad \text{for } i = 1, \ldots, k.$$

To simulate this uniform distribution is of course easy to do directly, but for the purpose of illustrating the sandwiching idea, we will insist on obtaining it using the Propp–Wilson algorithm for the ladder walk.

We obtain a valid update function for the ladder walk on k vertices by applying (21), which yields

$$\phi(1, x) = \begin{cases} 1 & \text{for } x \in [0, \frac{1}{2}) \\ 2 & \text{for } x \in [\frac{1}{2}, 1], \end{cases} \tag{85}$$

$$\phi(k, x) = \begin{cases} k-1 & \text{for } x \in [0, \frac{1}{2}) \\ k & \text{for } x \in [\frac{1}{2}, 1], \end{cases} \tag{86}$$

and, for $i = 2, \ldots, k-1$,

$$\phi(i, x) = \begin{cases} i-1 & \text{for } x \in [0, \frac{1}{2}) \\ i+1 & \text{for } x \in [\frac{1}{2}, 1]. \end{cases} \tag{87}$$

This update function can informally be described as follows: if $x < \frac{1}{2}$, then try to take a step down on the ladder, while if $x \geq \frac{1}{2}$, then try to take a step up.

Consider now the standard Propp–Wilson algorithm (as introduced in the previous chapter) for this Markov chain, with update function ϕ as in (85)–(87), and negative starting times $(N_1, N_2, \ldots) = (1, 2, 4, 8, \ldots)$. A typical run of the algorithm for the ladder walk with $k = 5$ is shown in Figure 11.

Note in Figure 11 that no two transitions "cross" each other, i.e., that a Markov chain starting in a higher state never dips below a chain starting at the same time in a lower state. This is because the update function defined in (85)–(87) preserves ordering between states, in the sense that for all $x \in [0, 1]$ and all $i, j \in \{1, \ldots, k\}$ such that $i \leq j$, we have

$$\phi(i, x) \leq \phi(j, x). \tag{88}$$

For a proof of this fact, see Problem 11.1.

It follows that any chain starting in some state $i \in \{2, \ldots, k-1\}$ always remains between the chain starting in state 1 and the chain starting in state k (this explains the term sandwiching). Hence, once the top and the bottom chains meet, all the chains starting in intermediate values have to join them as well; see, e.g., the realizations starting from time -8 in Figure 11. In order

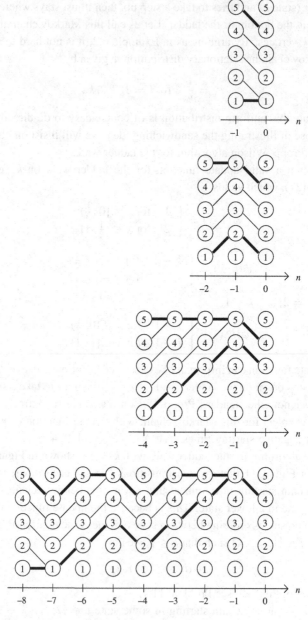

Fig. 11. A run of the Propp–Wilson algorithm for the ladder walk with $k = 5$. This particular run resulted in coalescence from starting time $-N_4 = -8$. Only those transitions that are actually carried out in the algorithm are drawn, while the others (corresponding to the dashed lines in Figure 8) are omitted. The chains starting from the top ($i = 5$) and bottom ($i = 1$) states are drawn in thick lines.

to check whether coalescence between all chains has taken place, we therefore only need to check whether the top and the bottom chains have met. But this in turn means that we do not even need to bother with running all the intermediate chains – running the top and bottom ones is enough! For the case $k = 5$ illustrated in Figure 11, this is perhaps not such a big deal, but for, say, $k = 10^3$ or $k = 10^6$, it is of course a substantial simplification to run just two chains rather than all k.

The next example to which we shall apply the sandwiching technique is the famous **Ising model**.

> **Example 11.1: The Ising model.** Let $G = (V, E)$ be a graph. The Ising model is a way of picking a random element of $\{-1, 1\}^V$, i.e., of randomly assigning -1's and $+1$'s to the vertices of G. The classical physical interpretation of the model is to think of the vertices as atoms in a ferromagnetic material, and of -1's and $+1$'s as two possible spin orientations of the atoms. Two quantities that determine the probability distributions have names taken from this physical interpretation: the **inverse temperature** $\beta \geq 0$, which is a fixed positive parameter of the model, and the **energy** $H(\xi)$ of a spin configuration $\xi \in \{-1, 1\}^V$ defined as
>
> $$H(\xi) = - \sum_{\langle x,y \rangle \in E} \xi(x)\xi(y). \qquad (89)$$
>
> Each edge adds 1 to the energy if its endpoints have opposite spins, and subtracts 1 otherwise. Hence, low energy of a configuration corresponds to a large amount of agreement between neighboring vertices. The Ising model on G at inverse temperature β means that we pick a random spin configuration $X \in \{-1, 1\}^V$ according to the probability measure $\pi_{G,\beta}$ which to each $\xi \in \{-1, 1\}^V$ assigns probability[40]
>
> $$\pi_{G,\beta}(\xi) = \frac{1}{Z_{G,\beta}} \exp\left(-\beta H(\xi)\right) = \frac{1}{Z_{G,\beta}} \exp\left(\beta \sum_{\langle x,y \rangle \in E} \xi(x)\xi(y) \right) \qquad (90)$$
>
> where $Z_{G,\beta} = \sum_{\eta \in \{-1,1\}^V} \exp(-\beta H(\eta))$ is a normalizing constant, making the probabilities of all $\xi \in \{-1, 1\}^V$ sum to 1. In the case $\beta = 0$ (infinite temperature), every spin configuration $\xi \in \{-1, 1\}^V$ has the same probability, so that each vertex independently takes the value -1 or $+1$ with probability $\frac{1}{2}$ each. If we take $\beta > 0$, the model favors configurations with low energy, i.e., those where most neighboring pairs of vertices take the same spin value. This effect becomes stronger the larger β is, and in the limit as $\beta \to \infty$ (zero temperature), the probability mass is divided equally between the "all plus" configuration and the "all minus" configuration. See Figure 12 for an example of how β influences the behavior of the model on a square lattice of size 15×15.

---

[40] The minus signs in (89) and in the expression $e^{-\beta H(\xi)}$ in (90) cancel each other, so it seems that it would be mathematically simpler to define energy differently by removing both minus signs. Physically, however, the present definition makes more sense, since nature tends to prefer states with low energy to ones with high energy.

From a physics point of view, the main reason why the Ising model is interesting is that it exhibits certain **phase transition** phenomena on various graph structures. This means that the model's behavior depends qualitatively on whether the parameter β is above or below a certain threshold value. For instance, consider the case of the square lattice of size $m \times m$. It turns out that if β is less than the so-called Onsager critical value[41] $\beta_c = \frac{1}{2}\log(1 + \sqrt{2}) \approx 0.441$, then the dependencies between spins are sufficiently weak for a Law of Large Numbers[42]-type result to hold: the proportion of $+1$ spins will tend to $\frac{1}{2}$ as $m \to \infty$. On the other hand, when $\beta > \beta_c$, the limiting behavior as m gets large is that one of the spins takes over and forms a vast majority. Some hints about this behavior can perhaps be read off from Figure 12. The physical interpretation of this phase transition phenomenon is that the ferromagnetic material is spontaneously magnetized at low but not at high temperatures.

We shall now go on to see how the Propp–Wilson algorithm combined with sandwiching applies to simulation of the Ising model. This particular example is worth studying for at least two reasons. Firstly, the Ising model has important applications (not only in physics but also in various other sciences as well as in image analysis and spatial statistics). Secondly, it is of some historical interest: it was to a large extent due to the impressive achievement of generating an exact sample from the Ising model at the critical value β_c on a square lattice of size 2100×2100 that the work of Propp & Wilson [PW] was so quickly recognized as seminal, and taken up by a large community of other researchers.

Before reading on, the reader is well-advised to try to obtain some additional understanding of the Ising model by solving Problem 11.3.

Example 11.2: Simulation algorithms for the Ising model. As a first step towards obtaining a Propp–Wilson algorithm for the Ising model, we first construct a Gibbs sampler for the model, which will then be used as a building block in the Propp–Wilson algorithm.

Consider the Ising model at inverse temperature β on a graph $G = (V, E)$ with k vertices. The Gibbs sampler for this model is a $\{-1, 1\}^V$-valued Markov chain $(X_0, X_1, \ldots)$ with evolution as follows (we will simply follow the Gibbs sampler recipe from Chapter 7). Given X_n, we obtain X_{n+1} by picking a vertex $x \in V$ at random, and picking $X_{n+1}(x)$ according to the conditional distribution (under the probability measure $\pi_{G,\beta}$) given the X_n-spins at all vertices except x, and leaving the spins at the latter set $V \setminus \{x\}$ of vertices unchanged. The updating of the chosen vertex v may be done using a random number U_{n+1} (as usual,

---

[41] Similar thresholds are known to exist for cubic and other lattices in 3 dimensions (and also in higher dimensions), but the exact values are not known.

[42] Theorem 1.2.

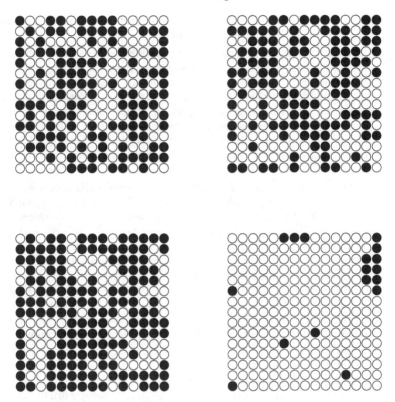

Fig. 12. Simulations of the Ising model on a 15×15 square lattice (vertical and horizontal nearest neighbors share edges), at parameter values $\beta = 0$ (upper left), $\beta = 0.15$ (upper right), $\beta = 0.3$ (lower left) and $\beta = 0.5$ (lower right). Black vertices represent $+1$'s, and white vertices represent -1's. In the case $\beta = 0$, the spins are i.i.d. Taking $\beta > 0$ means favoring agreement between neighbors, leading to clumping of like spins. In the case $\beta = 0.15$, the clumping is just barely noticable compared to the i.i.d. case, while already $\beta = 0.3$ appears to disrupt the balance between $+1$'s and -1's. This unbalance is even more marked when β is raised to 0.5. The fact that the fourth simulation ($\beta = 0.5$) resulted in a majority of -1's (rather than $+1$'s) is just a coincidence; the model is symmetric with respect to interchange of -1's and $+1$'s, so we were equally likely to get a similar majority of $+1$'s.

uniformly distributed on $[0, 1]$), and setting

$$X_{n+1}(x) = \begin{cases} +1 & \text{if } U_{n+1} < \frac{\exp(2\beta(k_+(x,\xi)-k_-(x,\xi)))}{\exp(2\beta(k_+(x,\xi)-k_-(x,\xi)))+1} \\ -1 & \text{otherwise,} \end{cases} \qquad (91)$$

where (as in Problem 11.3) $k_+(x, X_n)$ denotes the number of neighbors of x having X_n-spin $+1$, and $k_-(x, X_n)$ similarly denotes the number of such vertices having X_n-spin -1. That (91) gives the desired conditional distribution of $X_{n+1}(x)$ follows from formula (95) in Problem 11.3 (b).

Let us now construct a Propp–Wilson algorithm based on this Gibbs sampler. In the original version of the algorithm (without sandwiching), we have 2^k Markov chains to run in parallel: one from each possible spin configuration $\xi \in \{-1, 1\}^V$. In running these chains, it seems most reasonable to pick (at each time n) the same vertex x to update in all the Markov chains, and also to use the same random number U_{n+1} in all of them when updating the spin at x according to (91). To fully specify the algorithm, the only thing that remains to decide is the sequence of starting times, and as usual we may take $(N_1, N_2, \ldots) = (1, 2, 4, 8, \ldots)$.

How can we apply the idea of sandwiching to simplify this algorithm? First of all, sandwiching requires that we have some ordering of the state space $S = \{-1, 1\}^V$. To this end, we shall use the same ordering $\preceq$ as in Problem 11.3 (c), meaning that for two configurations $\xi, \eta \in \{-1, 1\}^V$, we write $\xi \preceq \eta$ if $\xi(x) \le \eta(x)$ for all $x \in V$. (Note that this ordering, unlike the one we used for the ladder walk, is not a so-called **total ordering** of the state space, because there are (many) choices of configurations ξ and η that are not ordered, i.e., we have neither $\xi \preceq \eta$ nor $\eta \preceq \xi$.) In this ordering, we have one **maximal** spin configuration ξ^{max} with the property that $\xi \preceq \xi^{max}$ for all $\xi \in \{-1, 1\}^V$, obtained by taking $\xi^{max}(x) = +1$ for all $x \in V$. Similarly, the configuration $\xi^{min} \in \{-1, 1\}^V$ obtained by setting $\xi^{min}(x) = -1$ for all $x \in V$ is the unique **minimal** spin configuration, satisfying $\xi^{min} \preceq \xi$ for all $\xi \in \{-1, 1\}^V$.

Consider now two of the 2^k different Markov chains run in parallel in the Propp–Wilson algorithm, starting at time $-N_j$: let us denote the two chains by $(X_{-N_j}, X_{-N_j+1}, \ldots, X_0)$ and $(X'_{-N_j}, X'_{-N_j+1}, \ldots)$. Suppose that the starting configurations X_{-N_j} and X'_{-N_j} satisfy $X_{-N_j}(x) \le X'_{-N_j}(x)$ for all $x \in V$, or in other words $X_{-N_j} \preceq X'_{-N_j}$. We claim that

$$X_{-N_j+1}(x) \le X'_{-N_j+1}(x) \tag{92}$$

for all $x \in V$, so that $X_{-N_j+1} \preceq X'_{-N_j+1}$. For any x other than the one chosen to be updated, this is obvious since $X_{-N_j+1}(x) = X_{-N_j}(x)$ and $X'_{-N_j+1}(x) = X'_{-N_j}(x)$. When x is the vertex chosen to be updated, (92) follows from (91) in combination with equation (96) in Problem 11.3 (c) (check this!). So we have just shown that $X_{-N_j} \preceq X'_{-N_j}$ implies $X_{-N_j+1} \preceq X'_{-N_j+1}$. By the same argument, $X_{-N_j+1} \preceq X'_{-N_j+1}$ implies $X_{-N_j+2} \preceq X'_{-N_j+2}$, and by iterating this argument, we have that

$$\text{if } X_{-N_j} \preceq X'_{-N_j} \text{ then } X_0 \preceq X'_0. \tag{93}$$

Now write $(X^{top}_{-N_j}, X^{top}_{-N_j+1}, \ldots, X^{top}_0)$ and $(X^{bottom}_{-N_j}, X^{bottom}_{-N_j+1}, \ldots, X^{bottom}_0)$ for the two chains starting in the extreme configurations $X^{top}_{-N_j} = \xi^{max}$ and

$X^{bottom}_{-N_j} = \xi^{min}$. As a special case of (93) we get

$$X^{bottom}_0 \preceq X_0 \preceq X^{top}_0$$

where $(X_{-N_j}, X_{-N_j+1}, \ldots, X_0)$ is any of the other $2^k - 2$ Markov chains. But now we can argue in the same way as for the sandwiching trick for the ladder walk: If the top chain $(X^{top}_{-N_j}, X^{top}_{-N_j+1}, \ldots, X^{top}_0)$ and the bottom chain $(X^{bottom}_{-N_j}, X^{bottom}_{-N_j+1}, \ldots, X^{bottom}_0)$ have coalesced by time 0, then all of the other $2^k - 2$ chains must have coalesced with them as well. So in order to check coalescence between all chains, it suffices to check it for the top and the bottom chain, and therefore the top and the bottom chains are the only ones we need to run! This reduces the task of running 2^k different chains in parallel to one of running just two chains. For large or even just moderately-sized graphs (such as those having, say, $k = 100$ vertices), this transforms the Propp–Wilson algorithm from being computationally completely hopeless, to something that actually works in practice.[43]

We shall not make any attempt here to determine in general to which Markov chains the sandwiching idea is applicable, and to which it is not. This has already been studied quite extensively in the literature; see Chapter 14 for some references. Problem 11.2 concerns this issue in the special case of birth-and-death processes.

Problems

11.1 **(5)** Show that the update function $\phi(i, x)$ for the ladder walk, defined in (85)–(87), is increasing in i. In other words, show that (88) holds for all $x \in [0, 1]$ and all $i, j \in \{1, \ldots, k\}$ such that $i \leq j$. Hint: consider the cases $x \in [0, \frac{1}{2})$ and $x \in [\frac{1}{2}, 1]$ separately.

11.2 **(9)** Note that the ladder walk is a special case of the birth-and-death processes defined in Example 6.2.

---

[43] The question of whether the algorithm works in practice is actually a little more complicated than this, because we need the top and the bottom chains to coalesce "within reasonable time", and whether or not this happens depends on G and on the parameter β. Take for instance the case of a square lattice of size $m \times m$ (so that $k = m^2$). It turns out that for β less than the Onsager critical value $\beta_c \approx 0.441$, the time to coalescence grows like a (low-degree) polynomial in m, whereas for $\beta > \beta_c$ it grows exponentially in m. Therefore, for large square lattices, the algorithm runs reasonably quickly when $\beta < \beta_c$, but takes an astronomical amount of time (and is therefore useless) when $\beta > \beta_c$. (This dichotomy is intimately related to the phase transition behavior discussed in Example 11.1.) As demonstrated by Propp & Wilson [PW], it is nevertheless possible to obtain exact samples from the Ising model on such graphs at large β by another ingenious trick, which involves applying the Propp–Wilson algorithm not directly to the Ising model, but to a certain graphical representation known as the Fortuin–Kasteleyn random-cluster model, and then translating the result to the Ising model.

(a) Can you find some useful sufficient condition on the transition probabilities of a birth-and-death process, which ensures that the same sandwiching idea as for the ladder walk will work?

(b) On the other hand, give an example of a birth-and-death process for which the sandwiching idea does *not* work.

11.3 (8) Consider the Ising model at inverse temperature β on a graph $G = (V, E)$. Let x be a particular vertex in V, and let $\xi \in \{-1, 1\}^{V \setminus \{x\}}$ be an arbitrary assignment of -1's and $+1$'s to the vertices of G *except for x*. Let $\xi^+ \in \{-1, 1\}^V$ be the spin configuration for G which agrees with ξ on $V \setminus \{x\}$ and which takes the value $+1$ at x. Similarly, define $\xi^- \in \{-1, 1\}^V$ to be the spin configuration for G which agrees with ξ on $V \setminus \{x\}$ and which takes the value -1 at x. Also define $k_+(x, \xi)$ to be the number of neighbors of x that take the value $+1$ in ξ, and analogously let $k_-(x, \xi)$ be the number of neighbors of x whose value in ξ is -1.

(a) Show that

$$\frac{\pi_{G,\beta}(\xi^+)}{\pi_{G,\beta}(\xi^-)} = \exp(2\beta(k_+(x, \xi) - k_-(x, \xi))). \tag{94}$$

Hint: use the definition (90), and demonstrate that almost everything cancels in the left-hand side of (94).

(b) Suppose that the random spin configuration $X \in \{-1, 1\}^V$ is chosen according to $\pi_{G,\beta}$. Imagine that we take a look at the spin configuration $X(V \setminus \{x\})$ but hide the spin $X(x)$, and discover that $X(V \setminus \{x\}) = \xi$. Now we are interested in the conditional distribution of the spin at x. Use (94) to show that

$$\pi_{G,\beta}(X(x) = +1 \mid X(V \setminus \{x\}) = \xi) = \frac{\exp(2\beta(k_+(x, \xi) - k_-(x, \xi)))}{\exp(2\beta(k_+(x, \xi) - k_-(x, \xi))) + 1} \tag{95}$$

holds,[44] for any $x \in V$ and any $\xi \in \{-1, 1\}^{V \setminus \{x\}}$.

(c) For two configurations $\xi, \eta \in \{-1, 1\}^{V \setminus \{x\}}$, we write $\xi \preceq \eta$ if $\xi(y) \leq \eta(y)$ for all $y \in V \setminus \{x\}$. Use (95) to show that if $\xi \preceq \eta$, then

$$\pi_{G,\beta}(X(x) = +1 \mid X(V \setminus \{x\}) = \xi) \leq \pi_{G,\beta}(X(x) = +1 \mid X(V \setminus \{x\}) = \eta). \tag{96}$$

11.4 (8*) Implement and run the Propp–Wilson algorithm for the Ising model as described in Example 11.2, on a square lattice of size $m \times m$ for various values of m and the inverse temperature parameter β. Note how the running time varies[45] with m and β.

---

[44] One particular consequence of (95) is that the conditional distribution of $X(x)$ given $X(V \setminus \{x\})$ depends only on the spins attained at the neighbors of x. This is somewhat analogous to the definition of a Markov chain, and is called the **Markov random field** property of the Ising model.

[45] In view of the discussion in Footnote 43, do not be surprised if the algorithm seems not to terminate at all for m large and β above the Onsager critical value $\beta_c \approx 0.441$.

12

Propp–Wilson with
read-once randomness

A drawback of the Propp–Wilson algorithm introduced in the previous two chapters is the need to reuse old random numbers: Recall that Markov chains are started at times $-N_1, -N_2, \ldots$ (where $N_1 < N_2 < \cdots$) and so on until j is large enough so that starting from time $-N_j$ gives coalescence at time 0. A crucial ingredient in the algorithm is that when the Markov chains start at time $-N_i$, the same random numbers as in previous runs should be used from time $-N_{i-1}$ and onwards. The typical implementation of the algorithm is therefore to store all new random numbers, and to read them again when needed in later runs. This may of course be costly in terms of computer memory, and the worst-case scenario is that one suddenly is forced to abort a simulation when the computer has run out of memory.

Various approaches to coping with this problem have been tried. For instance, some practitioners of the algorithm have circumvented the need for storage of random numbers by certain manipulations of (the seeds of) the random number generator. Such manipulations may, however, lead to all kinds of unexpected and unpleasant problems, and we therefore advise the reader to avoid them.

There have also been various attempts to modify the Propp–Wilson algorithm in such a way that each random number only needs to be used once. For instance, one could modify the algorithm by using new random variables each time that old ones are supposed to be used. Unfortunately, as we saw in Example 10.2, this approach leads to the output not having the desired distribution, and is therefore useless. Another common suggestion is to run the Markov chains not *from the past* until time 0, but from time 0 *into the future* until coalescence takes place. This, however, also leads in general to an output with the wrong distribution, as seen in Example 10.1.

The first satisfactory modification of the Propp–Wilson algorithm avoiding storage and reuse of random numbers was recently obtained by David

Wilson himself, in [W1]. His new scheme, which we have decided to call **Wilson's modification** of the Propp–Wilson algorithm, is a kind of coupling *into the future* procedure, but unlike in Example 10.1, we don't stop as soon as coalescence has been reached, but continue for a certain (random) extra amount of time. This extra amount of time is obtained in a somewhat involved manner. The remainder of this chapter will be devoted to an attempt at a precise description of Wilson's modification, together with an explanation of why it produces a correct (unbiased) sample from the stationary distribution of the Markov chain.

Although Wilson's modification runs *into the future*, it is easier to understand it if we first consider some variations of the *from the past* procedure in Chapter 10, and this is what we will do.

To begin with, note that although in Chapter 10 we focused mainly on starting times of the Markov chains given by $(N_1, N_2, \ldots) = (1, 2, 4, 8, \ldots)$, any strictly increasing sequence of positive integers will work just as well (this is clear from the proof of Theorem 10.1).

Next, let $N_1 < N_2 < \cdots$ be a *random* strictly increasing sequence of positive integers, and take it to be independent of the random variables $U_0, U_{-1}, U_{-2}, \ldots$ used in the Propp–Wilson algorithm.[46] Then the Propp–Wilson algorithm with starting times $-N_1, -N_2, \ldots$ still produces unbiased samples from the target distribution. This is most easily seen by conditioning on the outcome of the random variables $N_1, N_2, \ldots$, and then using the proof of Theorem 10.1 to see that, given $(N_1, N_2, \ldots)$, the conditional distribution of the output still has the right distribution, and since this holds for any outcome of $(N_1, N_2, \ldots)$, the algorithm will produce an output with the correct distribution.

Furthermore, we note that there is no harm (except for the added running time) in continuing to run the chains from a few more earlier starting times $-N_i$ after coalescence at time 0 has been observed. This is because the chains will keep producing the same value at time 0.

Our next step will be to specify more precisely how to choose the random sequence $(N_1, N_2, \ldots)$. Let

$$N_1 = N_1^*$$
$$N_2 = N_1^* + N_2^*$$
$$N_3 = N_1^* + N_2^* + N_3^*$$
$$\vdots \qquad \vdots$$

---

[46] It is in fact even possible to dispense with this independence requirement, but we do not need this.

where $(N_1^*, N_2^*, \ldots)$ is an i.i.d. sequence of positive integer-valued random variables. We take the distribution of the N_i^*'s to be the same as the distribution of the time N needed to get coalescence in the coupling *into the future* algorithm of Example 10.1. The easiest way to generate the N_i^*-variables is to run chains as in Example 10.1 (independently for each i), and to take N_i^* to be the time taken to coalescence.

Now comes a key observation: We **claim** that

> the probability that the Propp–Wilson algorithm starting from the first starting time $-N_1 = -N_1^*$ results in coalescence by time 0 (so that no earlier starting times are needed) is at least $\frac{1}{2}$.

To see this, let M_1 denote the number of steps needed to get coalescence in the Propp–Wilson algorithm starting at time $-N_1$ (and running past time 0 if necessary). Then M_1 and N_1^* clearly have the same distribution, and since they are also independent we get (by symmetry) that

$$\mathbf{P}(M_1 \leq N_1^*) = \mathbf{P}(M_1 \geq N_1^*) \tag{97}$$

Note also that

$$
\begin{aligned}
\mathbf{P}(M_1 \leq N_1^*) + \mathbf{P}(M_1 \geq N_1^*) &= 1 - \mathbf{P}(M_1 > N_1^*) + 1 - \mathbf{P}(M_1 < N_1^*) \\
&= 2 - (\mathbf{P}(M_1 > N_1^*) + \mathbf{P}(M_1 < N_1^*)) \\
&= 2 - \mathbf{P}(M_1 \neq N_1^*) \\
&\geq 2 - 1 = 1. \tag{98}
\end{aligned}
$$

Combining (97) and (98), we get that $\mathbf{P}(M_1 \leq N_1^*) \geq \frac{1}{2}$, proving the above claim.

By similar reasoning, if we fail to get coalescence of the Propp–Wilson algorithm starting from time $-N_1$, then we have conditional probability at least $\frac{1}{2}$ for the event that the Propp–Wilson chains starting at time $-N_2 = -(N_1^* + N_2^*)$ coalesce no later than time $-N_1$. More generally, we have that given that we come as far as running the Propp–Wilson chains from time $-N_j = -(N_1^* + \cdots + N_j^*)$, we have conditional probability at least $\frac{1}{2}$ of getting coalescence before time $-N_{j-1}$. We call the jth restart of the Propp–Wilson algorithm **successful** if it results in a coalescence no later than time $-N_{j-1}$. Then each restart has (conditional on the previously carried out restarts) probability at least $\frac{1}{2}$ of being successful.

Let M_j denote the amount of time needed to get coalescence starting from time $-N_j$ in the Propp–Wilson algorithm. Note that the only thing that makes the probability of a successful restart not *equal* to $\frac{1}{2}$ is the possibility of getting a tie, $M_j = N_j^*$; this is clear from the calculation leading to (98).

Now, to simplify things, we prefer to work with a probability which is exactly $\frac{1}{2}$, rather than some unknown probability above $\frac{1}{2}$. To this end, we toss a fair coin (or, rather, simulate a fair coin toss) whenever a tie $M_j = N_j^*$ occurs, and declare the jth result to be *-**successful** if either

$$M_j < N_j^*$$

or

$M_j = N_j^*$ and the corresponding coin toss comes up heads

(so that in other words the coin toss acts as a tie-breaker). Then, clearly, each restart has probability exactly $\frac{1}{2}$ of being *-successful.

Our preliminary (and correct, but admittedly somewhat strange) variant of the Propp–Wilson algorithm is now to generate the starting times $-N_1, -N_2, \ldots$ as above, and to keep on until a restart is *-successful.

The next step will be to translate this variant into an algorithm with read-once randomness. For this, we need to understand the distribution of the number of *-**failing** (defined as the opposite of *-successful) restarts needed before getting a *-successful restart in the above algorithm. To do this, we pick up one of the standard items from the probabilist's (or gambler's) toolbox:

> **Example 12.1: The geometric distribution.** Fix $p \in (0, 1)$. An integer-valued random variable Y is said to be geometrically distributed with parameter p, if
>
> $$\mathbf{P}(Y = n) = p(1 - p)^n$$
>
> for $n = 0, 1, 2, \ldots$. Note that if we have a coin with heads-probability p which we toss repeatedly (and independently) until it comes up heads, then the number of tails we see is geometrically distributed with parameter p.

The number of *-failing restarts is clearly seen to be a geometrically distributed random variable with parameter $\frac{1}{2}$; let us denote it by Y. The final (and *-successful) restart thus takes place at time $-N_{Y+1}$ (because there are Y *-failing restarts, and one *-successful).

The key to Wilson's modification with read-once randomness is that we will find a way to *first* run the chains from time $-N_{Y+1}$ to time $-N_Y$, *then* from time $-N_Y$ to time $-N_{Y-1}$ and so on up to time 0 (without any prior attempts with starting times that fail to give coalescence at time 0).

To see how this is done, imagine running two independent copies of the coupling *into the future* algorithm in Example 10.1. We run both copies for the number of steps needed to give *both* copies coalescence; hence one of them may continue for some more steps after its own coalescence. Let us declare the copy which coalesces first to be the **winner**, and the other to be the **loser**, with the usual fair coin toss as the tie-breaker in case they coalesce simultaneously.

We call the procedure of running two copies of the *into the future* algorithm in this way a **twin run**.

A crucial observation now is that the evolution of the Markov chain from time $-N_{Y+1}$ to time $-N_Y$ has exactly the same distribution as the evolution of the winner of a twin run as above (this probably requires a few moments[47] of thought by the reader!). So the evolution of the Markov chains in the Propp–Wilson algorithm from time $-N_{Y+1}$ to time $-N_Y$ (at which coalescence has taken place) can be simulated using a twin run.

Next, we simulate a geometric $(\frac{1}{2})$ random variable Y to determine the number of $*$-failing restarts in the Propp–Wilson algorithm.

If we are lucky enough so that Y happens to be 0, then $-N_Y = 0$ (and we have coalescence at that time) then we are done: we have our sample from the stationary distribution of the Markov chain.

If, on the other hand, $Y \geq 1$, then we need to simulate the evolution of the Markov chain from time $-N_Y$ to time 0. The value of $X(-N_Y)$ has already been established using the first twin run. To simulate the evolution from time $-N_Y$ to time $-N_{Y-1}$, we may do another twin run, and let the chain evolve as in the *loser* of this twin run, where the loser runs from time 0 until the time at which the *winner* gets coalescence. This gives precisely the right distribution of the evolution $(X(-N_Y), X(-N_Y + 1), \ldots, X(-N_{Y-1}))$ from time $-N_Y$ to time $-N_{Y-1}$; to see this requires a few more moments[48] of thought. We then go on to simulate the chain from time $-N_{Y-1}$ to time $-N_{Y-2}$ in the same way using another twin run, and so on up to time 0.

The value of the chain at time 0 then has exactly the same distribution as the output of the Propp–Wilson algorithm described above. Hence it is a correct (unbiased) sample from the stationary distribution, and we did not have to store or reread any of the random numbers. This, dear reader, is Wilson's modification!

Problems

12.1 (5) Let Ω denote the set of all possible evolutions when running the coupling *into the future* algorithm in Example 10.1 (for some fixed Markov chain and update function). Consider running two independent copies A and B of that coupling *into the future* algorithm, and write X_A and X_B for the evolutions of the two copies (so that X_A and X_B are independent Ω-valued random variables). Declare the copy which coalesces first to be the winner, and the other copy to be the loser,

---

[47] This is standard mathematical jargon for something that may sometimes take rather longer. In any case, Problem 12.1 is designed to help you understand this point.

[48] See Footnote 47.

with a fair coin toss as tie-breaker. Write X_{winner} and X_{loser} for the evolutions of the winner and the loser, respectively.

(a) Show that

$$\mathbf{P}(X_A = \omega, X_B = \omega') = \mathbf{P}(X_A = \omega', X_B = \omega)$$

for any $\omega, \omega' \in \Omega$.

(b) Show that for any $\omega \in \Omega$, the events $\{X_{winner} = \omega\}$ and $\{A \text{ is the winner}\}$ are independent.

(c) Show that the distribution of X_{winner} is the same as the conditional distribution of X_A given the event $\{A \text{ is the winner}\}$.

12.2 (3) Show that a random variable X whose distribution is geometric with parameter $p \in (0, 1)$ has expectation $\mathbf{E}[X] = \frac{1}{p} - 1$.

12.3 (9) Use the result in Problem 12.2 to compare the expected running times in the original Propp–Wilson algorithm (with $(N_1, N_2, \ldots) = (1, 2, 4, 8, \ldots)$), and in Wilson's modification. In particular, show that the expected running times are of the same order of magnitude, in the sense that there exists a universal constant C such that the expected running time of one of the algorithms is no more than C times the other's.

13

Simulated annealing

The general problem considered in this chapter is the following. We have a set $S = \{s_1, \ldots, s_k\}$ and a function $f : S \to \mathbf{R}$. The objective is to find an $s_i \in S$ which minimizes (or, sometimes, maximizes) $f(s_i)$.

When the size k of S is small, then this problem is of course totally trivial – just compute $f(s_i)$ for $i = 1, \ldots, k$ and keep track sequentially of the smallest value so far, and for which s_i it was attained. What we should have in mind is the case where k is huge, so that this simple method becomes computationally too heavy to be useful in practice. Here are two examples.

Example 13.1: Optimal packing. Let G be a graph with vertex set V and edge set E. Suppose that we want to pack objects at the vertices of this graph, in such a way that

  (i) at most one object can be placed at each vertex, and
  (ii) no two objects can occupy adjacent vertices,

and that we want to squeeze in as many objects as possible under these constraints. If we represent objects by 1's and empty vertices by 0's, then, in the terminology of Example 7.1 (the hard-core model), the problem is to find (one of) the feasible[49] configuration(s) $\xi \in \{0, 1\}^V$ which maximizes the number of 1's.[50] As discussed in Example 7.1, the number of feasible configurations grows very quickly (exponentially) in the size of the graph, so that the above method of simply computing $f(\xi)$ (where in this case $f(\xi)$ is the number of 1's in ξ) for all ξ is practically impossible even for moderately large graphs.

Example 13.2: The travelling salesman problem. Suppose that we are given m cities, and a symmetric $m \times m$ matrix D with positive entries representing the distances between the cities. Imagine a salesman living in one of the cities,

---

[49] Recall that a configuration $\xi \in \{0, 1\}^V$ is said to be feasible if no two adjacent vertices are assigned value 1.

[50] For the 8×8 square grid in Figure 7, the optimal packing problem is trivial. Imagine the vertices as the squares of a chessboard, and place 1's at each of the 32 dark squares. This is easily seen to be optimal. But for other graph structures it may not be so easy to find an optimal packing.

needing to visit the other $m - 1$ cities and then to return home. In which order should he visit the cities in order to minimize the total distance travelled? This is equivalent to finding a permutation $\xi = (\xi_1, \ldots, \xi_m)$ of the set $(1, \ldots, m)$ which minimizes

$$f(\xi) = \sum_{i=1}^{m-1} D_{\xi_i, \xi_{i+1}} + D_{\xi_m, \xi_1} . \tag{99}$$

Again, the simple method of computing $f(\xi)$ for all ξ is useless unless the size of the problem (measured in the number of cities m) is very small, because the number of permutations ξ is $m!$, which grows (even faster than) exponentially in m.

A large number of methods for solving these kinds of optimization problems have been tried. Here we shall focus on one such method: **simulated annealing**.

The idea of simulated annealing is the following. Suppose that we run a Markov chain with state space S whose unique stationary distribution places most of its probability on states $s \in S$ with a small value of $f(s)$. If we run the chain for a sufficiently long time, then we are likely to end up in such a state s. Suppose now that we switch to running another Markov chain whose unique stationary distribution concentrates even more of its probability on states s that minimize $f(s)$, so that after a while we are even more likely to be in an f-minimizing state s. Then switch to a Markov chain with an even stronger preference for states that minimize f, and so on. It seems reasonable to hope that if this scheme is constructed with some care, then the probability of being in an f-minimizing state s at time n tends to 1 as $n \to \infty$.

If the first Markov chain has transition matrix P' and is run for time N_1, the second Markov chain has transition matrix P'' and is run for time N_2, and so on, then the whole algorithm can be viewed as an inhomogeneous Markov chain (recall Definition 2.2) with transition matrices

$$P^{(n)} = \begin{cases} P' & \text{for } n = 1, \ldots, N_1 \\ P'' & \text{for } n = N_1 + 1, \ldots, N_1 + N_2, \\ \vdots & \vdots \end{cases}$$

There is a general way to choose a probability distribution on S which puts most of its probability mass on states with a small value of s, namely to take a so-called **Boltzmann distribution**, defined below. A Markov chain with the Boltzmann distribution as its unique stationary distribution can then be constructed using the MCMC ideas in Chapter 7.

Definition 13.1 *The* **Boltzmann distribution** $\pi_{f,T}$ *on the finite set S, with*

energy function $f : S \to \mathbf{R}$ *and* **temperature parameter** $T > 0$, *is the probability distribution on S which to each element $s \in S$ assigns probability*

$$\pi_{f,T}(s) = \frac{1}{Z_{f,T}} \exp\left(\frac{-f(s)}{T}\right). \tag{100}$$

Here

$$Z_{f,T} = \sum_{s \in S} \exp\left(\frac{-f(s)}{T}\right) \tag{101}$$

is a normalizing constant ensuring that $\sum_{s \in S} \pi_{f,T}(s) = 1$.

Note that the factor $\frac{1}{T}$ plays exactly the same role as the inverse temperature parameter β does in the Ising model (Example 11.1). We mention also that when the goal is to maximize rather than to minimize f, it is useful to replace the Boltzmann distribution by the **modified Boltzmann distribution**, in which the exponent in (100) and (101) is $\frac{f(s)}{T}$ instead of $\frac{-f(s)}{T}$.

The following result tells us that the Boltzmann distribution with a small value of the temperature parameter T has the desired property of placing most of its probability on elements s that minimize $f(s)$.

Theorem 13.1 *Let S be a finite set and let $f : S \to \mathbf{R}$ be arbitrary. For $T > 0$, let $\alpha(T)$ denote the probability that a random element Y chosen according to the Boltzmann distribution $\pi_{f,T}$ on S satisfies*

$$f(Y) = \min_{s \in S} f(s).$$

Then

$$\lim_{T \to 0} \alpha(T) = 1.$$

Sketch proof We consider only the case where S has a unique f-minimizer; the case of several f-minimizers is left to Problem 13.1. Write (as usual) k for the number of elements of S. Let s be the unique f-minimizer, let $a = f(s)$ and let $b = \min_{s' \in S \setminus \{s\}} f(s')$. Note that $a < b$, so that

$$\lim_{T \to 0} \exp\left(\frac{a - b}{T}\right) = 0. \tag{102}$$

We get

$$\pi_{f,T}(s) = \frac{1}{Z_{f,T}} \exp\left(\frac{-a}{T}\right) = \frac{\exp\left(\frac{-a}{T}\right)}{\sum_{s' \in S} \exp\left(\frac{-f(s')}{T}\right)}$$

$$= \frac{\exp\left(\frac{-a}{T}\right)}{\exp\left(\frac{-a}{T}\right) + \sum_{s' \in S \setminus \{s\}} \exp\left(\frac{-f(s')}{T}\right)}$$

$$\geq \frac{\exp\left(\frac{-a}{T}\right)}{\exp\left(\frac{-a}{T}\right) + (k-1)\exp\left(\frac{-b}{T}\right)}$$

$$= \frac{1}{1 + (k-1)\exp\left(\frac{a-b}{T}\right)},$$

which tends to 1 as $T \to 0$, because of (102). Hence

$$\lim_{T \to 0} \pi_{f,T}(s) = 1,$$

as desired.   □

The design of a simulated annealing algorithm for finding an element $s \in S$ which minimizes $f(s)$ can now be carried out as follows. First construct an MCMC chain for simulating the Boltzmann distribution $\pi_{f,T}$ on S, with a general choice of T. Very often, this is done by constructing a Metropolis chain as indicated in the final part of Chapter 7. Then we fix a decreasing sequence of temperatures $T_1 > T_2 > T_3 > \cdots$ with T_i tending to 0 as $i \to \infty$ (hence the term **annealing**), and a sequence of positive integers $N_1, N_2, \ldots$. Starting from an arbitrary initial state in S, we run the chain at temperature T_1 for N_1 units of time, then at temperature T_2 for N_2 units of time, and so on.

The choice of $(T_1, T_2, \ldots)$ and $(N_1, N_2, \ldots)$ is called the annealing (or cooling) schedule, and is of crucial importance: How fast should the temperature tend to 0 as time $n \to \infty$? There exist theorems stating that if the temperature approaches 0 sufficiently slowly (which, e.g., can be accomplished by letting the sequence $(N_1, N_2, \ldots)$ grow sufficiently fast), then the probability of seeing an f-minimizer at time n does tend to 1 as $n \to \infty$.[51] The meaning of "sufficiently slowly" of course depends on the particular application. Unfortunately, the annealing schedules for which these theorems guarantee such convergence are in most cases so slow that we have to wait for an astronomical amount of time before having a temperature that is low enough that we can be anywhere near certain of having found an f-minimizer. Therefore, most annealing schedules in practical applications are faster than those for which

---

[51] One such theorem was proved by Geman & Geman [GG]: If the temperature $T^{(n)}$ at time n converges to 0 slowly enough so that

$$T^{(n)} \geq \frac{k(\max_{s \in S} f(s) - \min_{s \in S} f(s))}{\log n}$$

for all sufficiently large n, then the probability of seeing an f-minimizer at time n converges to 1 as $n \to \infty$.

the desired convergence is rigorously known. The danger of this is that if the cooling takes place *too* rapidly, then the Markov chain risks getting stuck in a local minimum, rather than in a global one; see Example 13.4 below. The choice of annealing schedule in practice is therefore a highly delicate balance: On the one hand, we want it to be fast enough to get convergence in reasonable time, and on the other hand, we want it to be slow enough to avoid converging to an element which is not an f-minimizer. This often requires quite a bit of experimentation, giving the method more the character of "engineering" than of "mathematics".

Example 13.3: Simulated annealing for the travelling salesman problem.
Consider the travelling salesman problem in Example 13.2. We wish to find the permutation $\xi = (\xi_1, \ldots, \xi_n)$ of $(1, \ldots, m)$ which minimizes the total distance $f(\xi)$ defined in (99). In order to get a simulated annealing algorithm for this problem, let us construct a Metropolis chain (see Chapter 7) for the Boltzmann distribution $\pi_{f,T}$ at temperature T on the set of permutations of $(1, \ldots, m)$. To this end, we first need to define which permutations to view as "neighbors", i.e., between which permutations to allow transitions in the Metropolis chain. A sensible choice is to declare two permutations ξ and ξ' to be neighbors if there exist $i, j \in \{1, \ldots, m\}$ with $i < j$ such that ξ' arises by "reversing" the segment $(\xi_i, \ldots, \xi_j)$, meaning that

$$
\begin{aligned}
\xi' = (\xi'_1, \ldots, \xi'_m) &= (\xi_1, \xi_2, \ldots, \xi_{i-1}, \xi_j, \xi_{j-1}, \ldots, \\
&\quad \xi_{i+1}, \xi_i, \xi_{j+1}, \xi_{j+2}, \ldots, \xi_m).
\end{aligned} \tag{103}
$$

This corresponds to removing two edges from the tour through all the cities, and inserting two other edges with the same four endpoints in such a way that a different tour is obtained; see Figure 13. The transition matrix for the Metropolis chain corresponding to this choice of neighborhood structure and the Boltzmann distribution at temperature T is obtained by inserting (100) into (46). We get

$$
P_{\xi, \xi'} = \begin{cases}
\frac{2}{m(m-1)} \min\left\{\exp\left(\frac{f(\xi) - f(\xi')}{T}\right), 1\right\} & \text{if } \xi \text{ and } \xi' \\
& \text{are neighbors} \\
0 & \text{if } \xi \neq \xi' \text{ are} \\
& \text{not neighbors} \\
1 - \sum_{\substack{\xi'' \\ \xi'' \sim \xi}} \frac{2}{m(m-1)} \min\left\{\exp\left(\frac{f(\xi) - f(\xi'')}{T}\right), 1\right\} & \text{if } \xi = \xi',
\end{cases}
$$

$$\tag{104}$$

where the sum is over all permutations ξ'' that are neighbors of ξ. This corresponds to the following transition mechanism. First pick $i, j \in \{1, \ldots, m\}$ uniformly from the set of all choices such that $i < j$. Then switch from the present permutation ξ to the permutation ξ' defined in (103) with probability $\min\left\{\exp\left(\frac{f(\xi) - f(\xi')}{T}\right), 1\right\}$, and stay in permutation ξ for another time unit with the remaining probability $1 - \min\left\{\exp\left(\frac{f(\xi) - f(\xi')}{T}\right), 1\right\}$. This chain has the

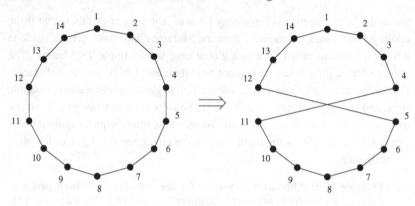

Fig. 13. A transition in the Metropolis chain in Example 13.3 for the travelling salesman problem, corresponding to going from the permutation $\xi = (1, 2, 3, 4, 5, 6, 7, 8, 9, 10, 11, 12, 13, 14)$ to the permutation $\xi' = (1, 2, 3, 4, 11, 10, 9, 8, 7, 6, 5, 12, 13, 14)$.

Boltzmann distribution $\pi_{f,T}$ as a reversible distribution, by the general Metropolis chain theory discussed in Chapter 7. The chain can also be shown to be irreducible (which is necessary in general for it to qualify as a useful MCMC chain).

It then only remains to decide upon a suitable cooling schedule, i.e., a suitable choice of $(T_1, T_2, \ldots)$ and $(N_1, N_2, \ldots)$ in the simulated annealing algorithm. Unfortunately, we have no better suggestion than to do this by trial and error.

We note one very happy circumstance of the above example: When inserting the Boltzmann distribution (100) into (46) to obtain (104), the normalizing constant $Z_{f,T}$ cancelled everywhere, because all the expressions involving the Boltzmann distribution were in fact ratios between Boltzmann probabilities. That is very good news, because otherwise we would have had to calculate $Z_{f,T}$, which is computationally infeasible. The same thing would happen for any Boltzmann distribution, and we therefore conclude that Metropolis chains are in general very convenient tools for simulating Boltzmann distributions.

Next, let us have a look at a simple example to warn against too rapid cooling schedules.

Example 13.4: The hazard of using a fast annealing schedule. Let $S = \{s_1, \ldots, s_4\}$, let $f : S \to \mathbf{R}$ be given by

$$\begin{cases} f(s_1) = 1 \\ f(s_2) = 2 \\ f(s_3) = 0 \\ f(s_4) = 2 \end{cases} \tag{105}$$

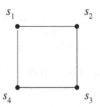

Fig. 14. The graph structure chosen for the Metropolis algorithm in Example 13.4.

and suppose that we want to find the minimum of $f(s_i)$ using simulated annealing.[52] To find a Metropolis chain for the Boltzmann distribution on S at temperature T, we need to impose a graph structure on S. Let's say that we opt for the square formation in Figure 14. By applying (100) and (46), the transition matrix

$$
\begin{bmatrix}
1 - e^{-1/T} & \frac{1}{2}e^{-1/T} & 0 & \frac{1}{2}e^{-1/T} \\
\frac{1}{2} & 0 & \frac{1}{2} & 0 \\
0 & \frac{1}{2}e^{-2/T} & 1 - e^{-2/T} & \frac{1}{2}e^{-2/T} \\
\frac{1}{2} & 0 & \frac{1}{2} & 0
\end{bmatrix}
$$

is obtained. Suppose now that we run the inhomogeneous Markov chain $(X_0, X_1, \ldots)$ on S, corresponding to some given annealing schedule, starting with $X_0 = s_1$. As in Footnote 51, write $T^{(n)}$ for the temperature at time n in this annealing schedule. Let A be the event that the chain remains in state s_1 forever (so that in particular the chain never finds the f-minimizing state s_3). We get

$$
\begin{aligned}
\mathbf{P}(A) &= \mathbf{P}(X_1 = s_1, X_2 = s_1, \ldots) \\
&= \lim_{n \to \infty} \mathbf{P}(X_1 = s_1, X_2 = s_1, \ldots, X_n = s_1) \\
&= \lim_{n \to \infty} \mathbf{P}(X_1 = s_1 \mid X_0 = s_1)\mathbf{P}(X_2 = s_1 \mid X_1 = s_1) \cdots \\
&\quad \times \mathbf{P}(X_n = s_1 \mid X_{n-1} = s_1) \\
&= \lim_{n \to \infty} \prod_{i=1}^{n} \left(1 - e^{-1/T^{(i)}}\right) = \prod_{i=1}^{\infty} \left(1 - e^{-1/T^{(i)}}\right)
\end{aligned}
$$

which is equal to 0 if and only if $\sum_{i=1}^{\infty} e^{-1/T^{(i)}} = \infty$. Hence, if $T^{(n)}$ is sent to 0 rapidly enough so that $\sum_{i=1}^{\infty} e^{-1/T^{(i)}} < \infty$, then $\mathbf{P}(A) > 0$, so that the chain may get stuck in state s_1 forever. This happens, e.g., if we take $T^{(n)} = \frac{1}{n}$. The simulated annealing algorithm then fails to find the true (global) f-minimizer f_3. Two factors combine to create this failure, namely

(i) the annealing schedule being too fast, and

---

[52] Of course, it is somewhat silly to use simulated annealing on a small problem like this one, where we can deduce that the minimum is $f(s_3) = 0$ by immediate inspection of (105). This example is chosen just to give the simplest possible illustration of a phenomenon that sometimes happens in simulated annealing algorithms for larger and more interesting problems.

(ii) state s_1 being a local f-minimizer (meaning that f takes a smaller value at s_1 than at any of the neighbors of s_1 in the graph structure used for the Metropolis chain) without being a global one.

Let us give one final example of an optimization problem for which simulated annealing may be a suitable method.

> **Example 13.5: Graph bisection.** Given a graph $G = (V, E)$ whose vertex set V contains $2k$ vertices, the graph bisection problem is to find a way of partitioning V into two sets V_1 and V_2 with k elements each, such that the total number of edges having one endpoint in V_1 and the other in V_2 is minimized. This problem is relevant for the design of search engines on the Internet: V may be the set of all web pages on which a given word was found, edges represent links from one page to another, and the hope is that V_1 and V_2 will provide a relevant split into different subareas.[53] For instance, if the search word is "football", then we may hope that V_1 contains mostly pages about American football, and V_2 mostly pages about soccer.

In recent years, several researchers have abandoned the idea of an annealing schedule, and instead preferred to run the Metropolis chain at a single fixed temperature, which is chosen on the basis of a careful mathematical analysis of the optimization problem at hand. For instance, Jerrum & Sorkin [JS] do this for the graph bisection problem in Example 13.5. They show that for k large and under reasonable assumptions on the input data, their algorithm finds, for arbitrary $\varepsilon > 0$, the optimal bisection in time $Ck^{2+\varepsilon}$ with overwhelming probability as $k \to \infty$, if T is taken to be of the order $n^{5/6+\varepsilon}$.

Problems

13.1 **(8)** Modify the proof of Theorem 13.1 in order to take care of the case where there are several elements $s \in S$ satisfying $f(s) = \min_{s' \in S} f(s')$.

13.2 **(6)** Describe a simulated annealing algorithm for the graph bisection problem in Example 13.5. In particular, suggest a natural choice of neighborhood structure in the underlying Metropolis chain.

13.3 **(8*)** Suppose that we want to solve the optimal packing problem in Example 13.1 (i.e., we want to maximize $f(\xi)$ over all feasible configurations $\xi \in \{0, 1\}^V$, where $f(\xi)$ is the number of 1's in ξ), and decide to try simulated annealing. To find suitable Markov chains, we start by considering Boltzmann distributions for the function $f(\xi)$. Since we are dealing with a maximization (rather than minimization) problem, we consider the modified Boltzmann distribution with the minus sign in the exponents of (100) and (101) removed.

(a) Show that this modified Boltzmann distribution at temperature T is the same as the probability measure $\mu_{G,\lambda}$ defined in Problem 7.4, with $\lambda = \exp\left(\frac{1}{T}\right)$.

---

[53] In this application, it is of course also natural to relax the requirement that V_1 and V_2 are of equal size.

(b) Implement and run a simulated annealing algorithm for some suitable instances of this problem. (Note that due to (a), the MCMC algorithm constructed in Problem 7.4 can be used for this purpose.)

14

Further reading

Markov theory is a huge subject (much bigger than indicated by these notes), and consequently there are many books written on it. Three books that have influenced the present text are the ones by Brémaud [B], Grimmett & Stirzaker [GS], and (the somewhat more advanced book by) Durrett [Du]. Another nice introduction to the topic is the book by Norris [N]. Some of my Swedish compatriots will perhaps prefer to consult the texts by Rydén & Lindgren [RL] and Enger & Grandell [EG]. The reader can find plenty of additional material (more general theory, as well as other directions for applications) in any of these references.

Still on the Markov theory side (Chapters 2–6) of this text, there are two particular topics that I would warmly recommend for further study to anyone with a taste for mathematical elegance and the power and simplicity of probabilistic arguments: The first one is **the coupling method**, which was used to prove Theorems 5.2 and 8.1, and which also underlies the algorithms in Chapters 10–12; see the books by Lindvall [L] and by Thorisson [T]. The second topic is the relation between **reversible Markov chains and electrical networks**, which is delightfully treated in the book by Doyle & Snell [DSn]. Häggström [H] gives a short introduction in Swedish.

Another goldmine for the ambitious student is the collection of papers edited by Snell [Sn], where many exciting topics in probability, several of which concern Markov chains and/or randomized algorithms, are presented on a level accessible to advanced undergraduates.

Moving on to the algorithmic side (Chapters 7–13), it is worth stressing again that the collection of algorithms considered here in no way is representative of the entire field of randomized algorithms. A reasonable overview can be obtained by reading, in addition to these notes, the book by Motwani & Raghavan [MR]. See also the recent collection edited by Habib & McDiarmid [HM] for more on randomized algorithms and other topics at the interface between probability and computer science.

The main standard reference for MCMC (Chapter 7) these days seems to be the the the book edited by Gilks, Richardson & Spiegelhalter [GRS]. Another book which is definitely worth reading is the research monograph by Sinclair [Si]. For the particular case of simulating the hard-core model described in Example 7.1, see, e.g., the paper by Luby & Vigoda [LV]. The problem discussed in Chapter 8 of proving fast convergence of Markov chains has been studied by many authors. Some key references in this area are Diaconis & Fill [DF], Diaconis & Strook [DSt], Sinclair [Si] and Randall & Tetali [RT]; see also the introductory paper by Rosenthal [R]. The treatment of q-colorings in Chapters 8 and 9 is based on the paper by Jerrum [J]. The general approach to counting in Chapter 9 is treated nicely in [Si].

Moving on to Propp–Wilson algorithms (Chapters 10–12), this is such a recent topic that it has not yet been treated in book form. The original 1996 paper by Propp & Wilson [PW] has already become a classic, and should be read by anyone wanting to dig deeper into this topic. Other papers that may serve as introductions to the Propp–Wilson algorithm are those by Häggström & Nelander [HN] and Dimakos [Di]. An annotated bibliography on the subject, continuously maintained by Wilson, can be found at the web site [W2]. For treatments of the sandwiching technique of Chapter 11, see [PW] or any of the other references mentioned here. The subtle issue of exactly under what conditions (on the Markov chain) the sandwiching technique is applicable is treated in a recent paper by Fill & Machida [FM]. The read-once variant of the Propp–Wilson algorithm considered in Chapter 12 was introduced by Wilson [W1].

For the purpose of refining MCMC methods in ways that lead to completely unbiased samples, there is an interesting alternative to the Propp–Wilson algorithm that has become known as **Fill's algorithm**. It was introduced by Fill [Fi], and then substantially generalized by Fill, Machida, Murdoch & Rosenthal [FMMR].

For an introduction to the Ising model considered in Chapter 11 (and also to some extent the hard-core model), see Georgii, Häggström & Maes [GHM].

Concerning simulated annealing (Chapter 13), see the contribution by B. Gidas to the aforementioned collection [Sn]. Also worthy of attention is the recent emphasis on running the algorithm at a carefully chosen fixed temperature; see Jerrum & Sorkin [JS].

Finally, let me emphasize once more that the difficult problem of making sure that we have access to a good (pseudo-)random number generator (as discussed very briefly in the beginning of Chapter 3) deserves serious attention. The classical reference for this problem is Knuth [K]. See also Goldreich [G] for an introduction to a promising new approach based on the theory of algorithmic complexity.

References

[B] Brémaud, P. (1998) *Markov Chains: Gibbs fields, Monte Carlo Simulation, and Queues*, Springer, New York.

[DF] Diaconis, P. & Fill, J. (1990) Strong stationary times via a new form of duality, *Annals of Probability* **18**, 483–522.

[DSt] Diaconis, P. & Strook, D. (1991) Geometric bounds for eigenvalues of Markov chains, *Annals of Applied Probability* **1**, 36–61.

[Di] Dimakos, X.K. (2001) A guide to exact simulation, *International Statistical Review* **69**, 27–48.

[DSn] Doyle, P. & Snell, J.L. (1984) *Random Walks and Electric Networks*, Mathematical Monographs 22, Mathematical Association of America.

[Du] Durrett, R. (1991) *Probability: Theory and Examples*, Wadsworth & Brooks/Cole, Pacific Grove.

[EG] Enger, J. & Grandell, J. (2000) *Markovprocesser och Köteori*, Mathematical Statistics, KTH, Stockholm.

[Fa] Fagin, R., Karlin, A., Kleinberg, J., Raghavan, P., Rajagopalan, S., Rubinfeld, R., Sudan, M. & Tomkins, A. (2000) Random walks with "back buttons", *Proceedings of the 32nd Annual ACM Symposium on Theory of Computing*, Portland, Oregon, pp. 484–493.

[Fi] Fill, J. (1998) An interruptible algorithm for perfect sampling via Markov chains, *Annals of Applied Probability* **8**, 131–162.

[FM] Fill, J. & Machida, M. (2001) Stochastic monotonicity and realizable monotonicity, *Annals of Probability*, **29**, 938–978.

[FMMR] Fill, J., Machida, M., Murdoch, D. & Rosenthal, J. (2000) Extension of Fill's perfect rejection sampling algorithm to general chains, *Random Structures and Algorithms* **17**, 290–316.

[GG] Geman, S. & Geman, D. (1984) Stochastic relaxation, Gibbs distributions, and the Bayesian restoration of images, *IEEE Transactions on Pattern Analysis and Machine Intelligence* **6**, 721–741.

[GHM] Georgii, H.-O., Häggström, O. & Maes, C. (2001) The random geometry of equilibrium phases, *Phase Transitions and Critical Phenomena, Volume 18* (C. Domb & J.L. Lebowitz, eds), pp. 1–142, Academic Press, London.

[GRS] Gilks, W., Richardson, S. & Spiegelhalter, D. (1996) *Markov Chain Monte Carlo in Practice*, Chapman & Hall, London.

[G] Goldreich, O. (1999) Pseudorandomness, *Notices of the American Mathematical Society* **46**, 1209–1216.

[GS] Grimmett, G. & Stirzaker, D. (1992) *Probability and Random Processes*, Clarendon, Oxford.

[HM] Habib, M. & McDiarmid, C. (2000) *Probabilistic Methods for Algorithmic Discrete Mathematics*, Springer, New York.

[H] Häggström, O. (2000) Slumpvandringar och likströmskretsar, *Elementa* **83**, 10–14.

[HN] Häggström, O. & Nelander, K. (1999) On exact simulation of Markov random fields using coupling from the past, *Scandinavian Journal of Statistics* **26**, 395–411.

[J] Jerrum, M. (1995) A very simple algorithm for estimating the number of k-colorings of a low-degree graph, *Random Structures and Algorithms* **7**, 157–165.

[JS] Jerrum, M. & Sorkin, G. (1998) The Metropolis algorithm for graph bisection, *Discrete Applied Mathematics* **82**, 155–175.

[K] Knuth, D. (1981) *The Art of Computer Programming. Volume 2. Seminumerical Algorithms*, 2nd edition, Addison–Wesley, Reading.

[L] Lindvall, T. (1992) *Lectures on the Coupling Method*, Wiley, New York.

[LV] Luby, M. & Vigoda, E. (1999) Fast convergence of the Glauber dynamics for sampling independent sets, *Random Structures and Algorithms* **15**, 229–241.

[MR] Motwani, R. & Raghavan, P. (1995) *Randomized Algorithms*, Cambridge University Press.

[N] Norris, J. (1997) *Markov Chains*, Cambridge University Press.

[PW] Propp J. & Wilson, D. (1996) Exact sampling with coupled Markov chains and applications to statistical mechanics, *Random Structures and Algorithms* **9**, 232–252.

[RT] Randall, D. & Tetali, P. (2000) Analyzing Glauber dynamics by comparison of Markov chains, *Journal of Mathematical Physics* **41**, 1598–1615.

[R] Rosenthal, J. (1995) Convergence rates for Markov chains, *SIAM Review* **37**, 387–405.

[RL] Rydén, T. & Lindgren, G. (1996) *Markovprocesser*, Lunds Universitet och Lunds Tekniska Högskola, Institutionen för matematisk statistik.

[Si] Sinclair, A. (1993) *Algorithms for Random Generation and Counting. A Markov Chain Approach*, Birkhäuser, Boston.

[Sn] Snell, J.L. (1994) *Topics in Contemporary Probability and its Applications*, CRC Press, Boca Raton.

[T] Thorisson, H. (2000) *Coupling, Stationarity, and Regeneration*, Springer, New York.

[W1] Wilson, D. (2000) How to couple from the past using a read-once source of randomness, *Random Structures and Algorithms* **16**, 85–113.

[W2] Wilson, D. (2001) Perfectly random sampling with Markov chains, `http://dimacs.rutgers.edu/~dbwilson/exact.html/`

Index

annealing schedule, 102, 104
aperiodicity, 25

Bernoulli (p) random variable, 6
binomial (n, p) random variable, 6, 43, 71
birth-and-death process, 41, 92
Boltzmann distribution, 100

Chebyshev's inequality, 6, 71, 74
chess, 27, 43
coin tossing, 8, 71, 96
conditional probability, 2
counting problem, 64
coupling, 34, 57, 62, 76, 81, 94

density, 3
distribution, 3

Ehrenfest's urn model, 43
equilibrium, 28, 34
expectation, 4

Fill's algorithm, 109
four-color theorem, 49

geometric distribution, 96
Gibbs sampler, 49
graph, 40
graph bisection, 106

hard-core model, 45, 47, 52, 99
hitting time, 29
homogeneity, 10

i.i.d. (independent and identically distributed),
    3
independence, 2
indicator function, 18

inhomogeneous Markov chain, 13, 50, 100
initial distribution, 10
initiation function, 18
Internet, 13, 52, 106
irreducibility, 23
Ising model, 87

ladder walk, 85
Law of Large Numbers, 7, 68, 69, 88

Markov chain, 10
Markov property, 9
MCMC (Markov chain Monte Carlo), 47
mean, 4
memoryless property, 9
Metropolis chain, 50

optimal packing, 99

perfect simulation, 76
phase transition, 88
polynomial time algorithm, 65
polynomial time approximation scheme, 66
probability measure, 1
Propp–Wilson algorithm, 76

random q-coloring, 49
random number generator, 17, 109
random variable, 2
random walk, 8, 40
reversibility, 39, 48, 51

simulated annealing, 100
St Petersburg paradox, 4
stationary distribution, 28, 34, 37, 39, 47
Steiner's formula, 5
systematic sweep Gibbs sampler, 50, 55

113

Printed in the United States
By Bookmasters